D1261353

Wissenschaftsethik und Technikfolgenbeurteilung
Band 9

Schriftenreihe der Europäischen Akademie zur Erforschung
von Folgen wissenschaftlich-technischer Entwicklungen
Bad Neuenahr-Ahrweiler GmbH
herausgegeben von Carl Friedrich Gethmann

Springer

Berlin
Heidelberg
New York
Barcelona
Hong Kong
London
Milan
Paris
Singapore
Tokyo

G. Banse, C. J. Langenbach, P. Machleidt (Eds.)

Towards
the Information Society

The Case of Central and Eastern European Countries

With 9 Figures and 16 Tables

 Springer

Reihenherausgeber
Professor Dr. Carl Friedrich Gethmann
Europäische Akademie GmbH
Wilhelmstraße 56, 53474 Bad Neuenahr-Ahrweiler, Germany

Bandherausgeber
Professor Dr. Gerhard Banse
ITAS, Forschungszentrum Karlsruhe
Postfach 3640, 76021 Karlsruhe, Germany

Dr.-Ing. Christian J. Langenbach
Europäische Akademie GmbH
Wilhelmstraße 56, 53474 Bad Neuenahr-Ahrweiler, Germany

Dr. Petr Machleidt
Centre of Science, Technology, Society Studies
at the Institute of Philosophy, Academy of Sciences
Jilská 1, CZ-110 00 Prague 1, Czech Republic

Redaktion
Dagmar Uhl, M.A.
Europäische Akademie GmbH
Wilhelmstraße 56, 53474 Bad Neuenahr-Ahrweiler, Germany

ISBN 3-540-41643-9 Springer-Verlag Berlin Heidelberg New York

CIP data applied for

Die Deutsche Bibliothek - CIP-Einheitsaufnahme
Banse, Gerhard: Towards the information society: the case of Eastern European countries /
Gerhard Banse; Christian J. Langenbach; Petr Machleidt. - Berlin; Heidelberg; New York;
Barcelona; Hong Kong; London; Milan; Paris; Singapore;
Tokyo: Springer, 2000
    (Wissenschaftsethik und Technikfolgenbeurteilung; Bd. 9)
    ISBN 3-540-41643-9

Springer-Verlag Berlin Heidelberg New York
a member of Bertelsmann Springer Science+Business Media GmbH

http://www.springer.de

Typesetting: Camera-ready by author
Coverlayout: de'blik, Berlin
SPIN: 10780369 Printed on acid-free paper 62/3020hu - 5 4 3 2 1 0 -

Europäische Akademie

zur Erforschung von Folgen wissenschaftlich-technischer Entwicklungen
Bad Neuenahr-Ahrweiler GmbH

Direktor:
Professor Dr. Carl Friedrich Gethmann

The Europäische Akademie

The *Europäische Akademie GmbH* is concerned with the scientific study of consequences of scientific and technological advance for the individual and social life and for the natural environment. The Europäische Akademie intends to contribute to a rational way of society of dealing with the consequences of scientific and technological developments. This aim is mainly realised in the development of recommendations for options to act, from the point of view of long-term societal acceptance. The work of the Europäische Akademie mostly takes place in temporary interdisciplinary project groups, whose members are recognised scientists from European universities. Overarching issues, e.g. from the fields of Technology Assessment or Ethics of Science, are dealt with by staff of the Europäische Akademie.

The Series

The Series "Ethics of Science and Technology Assessment" serves to publish the results of the Europäische Akademie's work. It is published by the Academy's director. Besides the final results of the project groups the series includes volumes on general questions of ethics of science and technology assessment as well as other monographic studies.

Foreword

Information and knowledge play an increasingly important role in the implementation of public policies, in particular those of the Central and Eastern European Countries. They are involved in many respects in the elaboration of scientific programs. They are more and more present in the political decision making process and as topic for scientific conferences. They are often at the centre of international discussions on related topics, for example, differences in approaches to produce and apply knowledge or different responses to social function of information.

A major lesson of these past years applies to democracy. Europeans demand more involvement in decisions that concern them. This demand goes well beyond decision making. For public action to be acceptable and efficient, the whole process should become more democratic, from the definition of the problems, to the implementation and the evaluation of solutions.

In the context of conducting research on the consequences of scientific and technological advance, the Europäische Akademie Bad Neuenahr-Ahrweiler in Germany and the Academy of Science of the Czech Republic organised a conference on the relationship between "democracy-participation-technology assessment" in February 1999. The objective of the workshop was to express and exchange various viewpoints and attitudes to the problems of transition "from information society to knowledge society."

The great response given to the international conference underlines the need not only for Central and Eastern European Countries to take into consideration more common projects like this for the future. Therefore the European Academy is also concerned with the support of scientists and scientific departments in Eastern Europe which are working within its research spectrum from the beginning. The idea came up to publish a volume where these matters could be shown within the scientific investigations of the project "TA-East" in the book series of the European Academy.

The book is addressed to researchers in the fields of social science, humanities, information technology and technology assessment in particular. It may be also of interest to policy-makers and the wider public concerned with information society. It was published in English in order to be accessible to an international audience.

July, 2000
Bad Neuenahr-Ahrweiler   Carl Friedrich Gethmann

Preface[1]

The subject of this working meeting of researchers in the social sciences "From Information Society to Knowledge Society: Democracy – Participation – Assessment of the Impact of Technology" is a very substantive theme indeed. It covers not only the technological and economic implications but also the social, political and ethical ramifications of the current developmental trend in our civilisation, affected by globalisation with its opportunities and challenges as well as risks and conflicts. Our society's tendency to develop towards a knowledge society may be perceived as a process in which we are in a position to apply its relatively high level of education and well-developed scientific base with its rich traditions, i.e., those vital attributes marking any modern developing society on the eve of a new millennium. Today it is not only the actual quantity of available capital and labour force but especially the level of knowledge which decides whether a nation or a state will eventually succeed in the global competition for economic and social prosperity. The Czech Republic does not have any appreciable resources of energy or raw materials. Science, research, technology and know-how, the use of the country's not small intellectual potential – all this points to the most logical and fastest way to prosperity.

This goal is also associated with the need for building a positive attitude in the general public towards scientific thinking, to science as an institution and as a way of perceiving the world. To participate in the creation of a positive picture of science in the public eye is a major task facing any scholar, whether in the natural or social sciences. At present, we are witnessing research projects paying priority attention to the study of real systems, projects marked by interdisciplinary links between the social, cultural, economic, technological and ecological dimensions of our society. There are signs that interdisciplinary efforts involving all experts in the exact, natural and social sciences should be intensified.

The international dimension of this conference is evident. I am pleased to see among its participants scholars from various types of institutions – universities, academic institutes, and state and private organisations from nine European countries. I believe that the absolute openness between the Academy on the other hand and Czech and foreign universities on the other is essential. It is gratifying to see that this conference is being attended by many scholars representing Czech and foreign universities and their research centres.

The high educational and cultural standards in the Czech Lands are the Czech Republic's contribution to a unifying Europe. Prague has always been a major cross-roads of European science and culture. This is corroborated by the presence

---

[1] Professor Zahradník's speech inaugurating the conference "From Information Society to Knowledge Society: Democracy – Participation – Assessment of the Impact of Technology"

here today of the distinguished European researchers engaged in the study of the social prerequisites and the impact of science and technological development. Indeed, this gives the Prague gathering an indisputably inspiring European dimension.

Together with the keynote of the conference, issues that pose a task of utmost importance to all the advanced countries – mainly due to its interdependence with the various technological, social, political, economic and ethical challenges – have now come to the forefront. Still, this task is felt particularly urgently in the countries which are, to a varying extent, in the grips of their difficult experience with social transformation.

This conference is jointly organised by two scientific institutions – a German and a Czech one. This can be seen as a major contribution to upgrading the quality of Czech-German relations. In this respect, science has, for years, been playing a positive and seminal role. We should continue cultivating our good neighbourly relations. The Academy of Sciences of the Czech Republic is known to maintain lively professional and friendly contacts with many research institutions in Germany.

Science is more than its institutional form, more than the sum-total of the research centres of the Academy of Sciences and the university institutions. First and foremost, it is a way of thinking and a method of information processing. It is an ability to assume a creative attitude to modern knowledge. Science also does not mean that we are creating a final, definitive and single description of reality. The growing breadth and depth of the available pool of knowledge is introducing an astonishing sense of order in certain domains, while disclosing, at the same time, many new and hitherto unexpected themes which offer what will undoubtedly prove to be an external adventure of learning.

I wish your Conference much success.

February, 1999
Prague   Rudolf Zahradník

Contents

III   The Application Fields of TA 93

IV   The Judgement 161

List of Authors..205

Outline

Gerhard Banse, Christian J. Langenbach, Petr Machleidt

A workshop of researchers with European dimensions in the field of social sciences took place in Prague from February 3 to February 5, 1999 titled "From an Information Society to a Knowledge Society: Democracy - Participation - Technology Assessment". This workshop was a common event of two institutions - it was organized by Europäische Akademie zur Erforschung von Folgen wissenschaftlich-technischer Entwicklungen, Bad Neuenahr-Ahrweiler GmbH in cooperation with the Centre for Science, Technology, and Society Studies at the Institute of Philosophy of the Academy of Sciences of the Czech Republic. More than 40 scientists and experts of various institutions from nine European countries - from universities, academic centers and from various governmental and private organizations - participated in the conference. Approximately two thirds of participants came from EU countries and one third from the Central and East European countries. Among others the following organizations took part in the event: Academy of Sciences of the Czech Republic, Charles University of Prague, Grant Agency of the Czech Republic, Czech Technical University of Prague, Czech Ministry of Defense, PIAS -Prague Institute of Advanced Studies, East Bohemian University of Pardubice, West Bohemian University of Plzeň, Matěj Bel University of Banská Bystrica, Silesian University of Katowice, University of Social Sciences of Tychy (Poland), Jagellonian University of Krakow, Academy of Sciences of Hungary and Technical University of Budapest, Academy of Sciences of Russia, International Institute for Global Problems (Moscow), ITAS - Institute for Technology Assessment and Systems Analysis (Karlsruhe), BSI – Federal Agency for Security of Information Technologies (Bonn), Brandenburg Technical University (Cottbus), University of Potsdam, University of Merseburg (Germany), Centre for Transfer of Technologies, GmbH Rheine (Germany), FES - Friedrich Ebert Foundation (Bonn), CSTS/CNAM - Centre of Science, Technology and Society Studies, (Paris), G. W. Rathenau[1] Institute (Holland), etc.

Given the structure of participants and the subject discussed, the Prague meeting was undoubtedly of European dimension. Among distinguished personages within the European scope who presented their papers were for example Professor J.-J. Salomon from France, Professor J. van Eijndhoven from

---

[1] Professor Dr. G. W. Rathenau (1911-1989) was one of the founders of Nederlandse Organisatie voor Technologisch Aspectenonderzoek (NOTA). NOTA was renamed to Rathenau Institute in 1994.

Holland, Professor C. F. Gethmann and Professor A. Grunwald from Germany and Professor L. Tondl, who is in the international expert community considered to be the initiator of social assessment of technical facilities and technical solutions in the Czech Republic.

Activities of workshop participants, relatively high number of countries represented as well as the diversity of participating institutions allowed to attain the established objectives of the conference. These were especially the following:

- to express and illustrate various viewpoints and attitudes to the problems of the presented subject matter
- to make use of the feedback effects, i.e. to attain the exchange of experience and information
- to consider possible common projects for the future

The conference was opened and the first evening session was presided over by the Chairman of the Academy of Sciences of the Czech Republic, Professor R. Zahradník. In the opening statement he classed the subject of the conference as extremely important since it dealt not only with diverse technological and economic, but also social, political and ethical consequences of the present civilization development. Directing the society towards the knowledge society cannot be understood only as some new theoretical construct – it is necessary to perceive it as a real change for transforming countries, including the Czech Republic, where it is possible to make use of their high level of education and developed traditional scientific base. Professor Zahradník assumed that the organization of the common conference with the above mentioned important subject would bring significant contributions to the improvement of the quality of reciprocal relations – and the contribution of science is irreplaceable here. The fact that this international conference was organized at the Academy of Sciences of the Czech Republic was considered by him to be a sign of intensive development of research cooperation among the two Academies and other participating institutions. At the same time, he expressed the hope that in future it would be possible to take up again the positive experience gathered during preparation of this conference.

Variety as well as hierarchical structure of communication links among specialists in the area of science, research and development and civil society were mentioned by Professor J.-J. Solomon of Center of Science, Technology and Society Studies, CNAM, Paris in his opening key address *"Science, Democracy and Technological Risks"*. He pointed out the need of clarity and transparency of resolving technological risks of these links and at the same time he emphasized the significance of democratic participation of the public and experts in the decision-making processes. Among other implications this assumes ensuring the quality of educational systems and their diversification.

The main thematic parts of the program were covered in the opening section of the conference by the Head of German institution Europäische Akademie zur Erforschung von Folgen wissenschaftlich-technischer Entwicklungen, Bad Neuenahr-Ahrweiler GmbH Professor C. F. Gethmann and by the Head of the Center for Science, Technology and Society Studies at the Institute of Philosophy of the Academy of Sciences of the Czech Republic, Professor L. Tondl. Both of them once again stressed the significance of the conference as an important turning point in the process of preparation and implementation of common projects that should provide means for contacts among the researchers in various parts of Europe – as a contribution to its integration.

The conference program was thematically and organizationally divided into three parts. The first of them concentrated on more general and conceptual considerations. The second section of the conference dealt with the application areas and the third one with the problems arising in the course of the evaluation process.

In the first part called "Conception" supervised by Professor G. Banse we listened to a contribution by G. Bechmann: *"The Idea and Concept of Information Society. Society as an Information Society"* and to a paper by S. Provazník, A. Filáček and P. Machleidt: *"Problems of the Production and Application of Knowledge in a Societal Transformation: The Czech Republic's Approach to Knowledge Society."* G. Bechmann based his work on the insight that possibilities to locate the structure changes allow to classify the social change of the society as the information society, the description of which may not be limited to technological processes only. New conceptions were introduced which analyze social changes initiated by information technologies: information society as information economy, as post-industrial society, as the end of industrial society of mass production and as a knowledge society. All conceptions are firmly rooted in an insufficiently defined concept of information. Thus Bechmann justifies his concept and the intention to explain the concept of information society by means of the concept of information, however, on the basis of the identification of differences between both the terms. Following discussion pointed to the need to stipulate the common approach to determination of the principles and problems of information society – in order to verify social, cultural and technological dimensions.

Common paper of the Czech authors analyzed the process of society transformation in the Czech Republic in terms of the knowledge society requirements. The paper emphasized the increasing significance of technological innovations and the need to make use of the existing resources, especially in the area of education and science in order to ensure development of the Czech Republic. It also demonstrated that in many respects there exist common transferable experience in various societies in transition, as shown in the papers of Czech and Hungarian participants on the conference. Complex transformation of the whole socio-political and economic system has not been and still is not trouble-free. This applies not only to significant reduction of the research base and especially applied research. It also concerns the problems of social, economic and value environment, which did change its institutional structure, but in many aspects continued to be influenced by persisting sets of values and criteria of industrial society without applying more advanced types of innovation initiatives. Research and development activities should take more progressive part in the creation of these activities in future development.

In section II of the conference called "Application Areas" conducted by Professor V. Gorokhov of Academy of Sciences of Russia, various views of the consequences of information technologies were presented. Professor I. Hronszky of TU Budapest referred to the modification of the technology change evaluation concept, in which the role and relevance of expertise change. This also includes increasing demands on expertise – on scientific approach, complexity and capability to predict risks and uncertainties of technological development. And from whence follows the question of the competence of experts and also the problem of participation of the public in the decision-making concerning technology development – this is considered by author to be a certain type of "social learning ". Some experience with using the Internet in Poland – e.g. risk that the possibility of Internet access will become the source of a new social inequality, exclusive character of its current way of use which is far from proclaimed democracy – were analyzed by Professor A. Kiepas of Silesian University Katowice. This was followed by the paper of Dr. O. Ulrich from BSI Bonn, who characterized the relevance of the information technologies safety not only from technological, but also from social and cultural point of view. In his opinion the safety of information technologies does not concern only the minority of current users of modern communication media, but the society as a whole, since the society increasingly faces the development of these technologies. It highlights the necessity of the development of "safety culture" and media competence. For example, with the digital signature law, as O. Ulrich emphasized, the era of new understanding of the term 'safety' is coming. Professor J. Vlček (Czech Technical University) presented the concept of information in terms of the systems theory. He outlined new possibilities and relevance of methods and techniques resulting from this perspective.

Third thematic part of the conference focused on various aspects of technology assessment. This section of the conference was supervised by Professor J. Eijndhoven, Rathenau Institute, Holland. Professor P. Tamas of Academy of Sciences of Hungary presented the analysis of information society development in Hungary. Movement of the society towards the information society in the "post-communist countries" takes place seemingly with no national strategies and without taking the cultural conditions of individual countries into consideration. Using the experience of Hungary he described the change of the paradigm of this process – the transition from political to economic premises. Professor A. Grunwald – Head of ITAS Karlsruhe – presented relevant and actual aspects of discussion in science, politics and in the public concerning the problems of knowledge. These were discussed using the concept of *evaluation and shaping technology*. The paper particularly stressed the aspects of rationality and participation as well as their significance for processes of social mastering of the adequate relation to technology. New demands on the concept of rationality which emphasize traditional technological and economic approaches and also social and humanity oriented criteria, considerations of prognostic dimensions and important aspects connected with responsibility – these are, in author's opinion, important topics of recent philosophy of technology.

Professor L. Tondl with his paper *"Knowledge and Value Prerequisites of Evaluation and Decision-Making"* brought the important area of human dimension and human values to the conference. He pointed out that what we call human dimensions, i.e. certain sets of aspects and criteria, includes various somatic, mental, as well as ethical aspects, the significance of which may differ considerably during various decision-making processes. Accelerating rhythm of innovation steps which started with discoveries and finished with technological applications had not always been accompanied with desirable changes in the area of value structures, life objectives and also in corresponding ethic areas. Until now the dominating value structure has been closely dependent on what formed the objectives, ambitions and preferences created during the expansion of industrial society. He stressed that in addition to the wide range of knowledge, the objectives of required education quality also included necessary value systems incorporating ethic, aesthetic and cultural dimensions. Necessary innovation activities which are the presumption of the knowledge society require new forms of communication processes. Professor Tondl suggested that practically in all areas of human activities, where results of contemporary science and technology are applied, it is necessary to utilize not only up-to-date knowledge, but also sets of values, criteria for decision-making, and for assessment of feasible alternatives and options.

In the final discussion headed by Professor G. Banse the initial assumption of conference organizers proved to be true - the problem under study requires international cooperation and interdisciplinary approach. As the key role of the knowledge in the scope of emerging trends towards the knowledge society we consider knowing the limitations of technical and technological applications, possible risks and potential failures that cannot be avoided. More and more important role here is played by the subjective factor, by human being in the full range of his capabilities and interests – this is what we call the human capital. This also includes the problems of participation and democracy which need to be further discussed with respect to the Internet and safety of information technologies. Searching for adequate relation to technology in the future assumes more comprehensive attitude to the public, which is much more an active bearer of information technologies development than its target.

The Prague Workshop became a vivid momentum for the further bilateral- and multilateral cooperation. The meeting provided a fair overview of the status and perspectives of TA in individual countries of Central and East Europe. However, it also showed the complexity of creation of space for TA type activities in individual countries – for independent activity of both experts and the public based on individual responsibility.

In principle, the book came from the contributions presented at the Prague Workshop. However, it reflects the ideas and incentives sounded in the discussion. Therefore, the book consists both of the final versions of the presented papers and the new contributions initiated by the Workshop.

We would like to thank Dagmar Uhl, M.A. for her significant contribution to the editing of the book. We thank our colleague, Dr. phil. Jiří Loudín, scientific member of the Center for Science, Technology and Society Studies at the Institute of Philosophy of the Academy of Sciences of the Czech Republic, for critical reviews of the different manuscripts and helpful suggestions for its improvement.

I Introduction

The Genesis of the Project „TA - East"

Gerhard Banse

1 Introduction

In Germany, knowledge about the practice of Technology Assessment (TA) and corresponding activities in East and East Central European nations is quite limited: in the relevant publications, there is only little – often outdated – information. This almost complete „lack of knowledge" concerns the political and legal prerequisites, as well as institutionalization, applies equally to projects as to literature. There are at least four reasons for this state of affairs:

1. There is quite a bit of prejudice in Germany as far as the realities and the possibilities in these countries are concerned. This often results in a lack of interest in the – often quite interesting – occurrences in these countries.
2. The development in East and East Central European nations is characterized by rapid change, by „ups" and „downs"; at first glance, however, no stable structures nor even tendencies are recognizable, not even to the West or Central European eye. The constellations of political power can shift, just as well as the fundamental economic conditions, scientific potentials can just as well be restructured and reorganized as national governments or regional administrations. Thereby, political objectives and priorities can also change (sometimes very quickly), as well as possibilities for social action and intervention. The continuity necessary for processes of consolidation and thorough differentiation is often lacking. For this reason, it is difficult to isolate generalizable facts in order to bring them to the attention of the West.
3. There are differences in terminological and conceptual respects. In Germany, as in France or Great Britain, quite heterogeneous scientific and political, methodological and ethical, participative as well as elitist, institutionalized as well as „free" concepts (which all certainly somehow have something in common), are concealed behind the abbreviation „TA" – if it is used at all. On the other hand, there are many and varied designations – in the German language area alone – from „Technology Impact Evaluation" via „Estimation" or „Judgement of Technological Impact" to „Technology Impact Assessment". Can one expect conceptual or terminological „assimilation" from people in the nations of East Central and Eastern Europe, who haven't had the chance to gain enough experience with the „Discussion Culture" prevalent here? For this reason, we find in

those countries some activities which in fact are TA, but are not called „TA", as well as some which are called „TA", but aren't.

4. Finally, language barriers also play a role: English or French are more familiar to us here in the West than Polish, Russian, Czech, Ukrainian or Hungarian, and the „TA-Datenbank Nachrichten" or the „EPTA Network Newsletter" are easier for us to read (and obtain) here than the magazines „Transformacje" („Transformation"/Polish), „Teory vedy" („Scientific Theory"/Czech), or „Kornyezet es felödes" („Environment and Development"/Hungarian). On the other hand, the „TA-Datenbank" and its advantages, for example, as well as the TAB (Technology Assessment Unit of the Bundestag) and its reports are often unknown east of the Oder River: in most places, the necessary information just isn't available.

In view of this situation, the Director of the Europäische Akademie Bad Neuenahr-Ahrweiler GmbH, Professor Gethmann, commissioned me in the summer of 1996 to develop a project, which would be able to help reduce these information deficits, and which the BMBF (Bundesministerium für Bildung und Forschung/Federal Ministry for Education and Research) would be willing to support. In accordance with this formulation of the assignment, the European Academy's project „Technology Assessment and the Ethics of Science in East Central European Nations. An Appraisal" – planned for a duration of two years – could take up its work in March 1997. The definition of its goals consisted primarily in evaluating information available and research directly „on the spot" in order to comprehend the situation in the field of Technology Assessment, the ethics of science and technology, and related questions concerning research and technology policy in three East Central European nations. The project was soon dubbed „TA East".

With the process of transformation in East Central and East European states in the past decade, the chances for progress in the fields of TA and the (practical) ethics of science and technology have, on the one hand, improved, because TA and the ethics of science as a means of advising politicians are – scientifically – much more widely recognized than was formerly the case, are supported by society, and are politically accepted; on the other hand, worsened, because the basic industrial as well as the financial conditions and the situation on the labor market in the individual nations show in general a negative trend. For this reason, the (financial) means for considerations which could – in the sense of political advice as decision support in social questions –

pave the way for mechanization, are probably quite limited (s. also the articles on the background theme „Science and Technology amid Change in Eastern Europe" in: Technology 1993).

For the purposes of our investigation, we can proceed on the basis of the following premises (s. also BMFT [Bundesministerium für Forschung und Technik = Federal Ministry for Research and Technology, now BMBF, s. above] 1991; Schmittel 1992):

(a) There are considerable environmental and economic problems caused by technology and by hazardous waste sites, especially in the fields of power generation, the chemical industry, agriculture, and transportation:

(b) Decisions have to be made on technical solutions which can modify, complement, or replace those previously employed on the one hand, and on technical solutions which have yet to be developed and put to use on the other;

(c) There is great need for general, synoptic knowledge and for a system of orientation as a basis for decisions on technology in politics, the economy, and science (in particular, when seen against the background of the reorganization of the entire industrial infrastructure).

(d) in public opinion, a process of sensitization has taken place, and consternation is expressed in view of the effects of technological developments and their implementation (also seen against the backdrop of once-denied opportunities for discussion and participation).

The project was assigned the task of improving the general state of information by means of a systematic and goal-directed inventory. This was concentrated intentionally on the East Central European countries Poland, the Czech Republic, and Hungary – in the first place, because these countries traditionally have held close scientific contact with Germany; in the second place, because the three nations chosen are the only East Central and East European countries (with the exception – e. g. – of some of the larger CIS-nations) which not only have the scientific potential, but have also reached a stage of technological development which makes this possible and necessary; in the third place, because these three states belong to those with the best prospects of being accepted as members in the EU in the near future.

As a result of systematic investigation during the project's initial phase (a number of interviews and discussions in Prague, Pardubice, and Budapest), we found reason to believe that the standard as well as the methodology of Technology Assessment would differ to a certain extent from the practice in Germany, but would also present novel and innovative developments (for example, industrial organizations and local governments as clients), a more comprehensive (holistic) view of things, conceptually new approaches, other forms of institutionalization. Simple wholesale imitation of models well-established in Germany (or in Central and Western Europe) – which originally actually was the case – failed, due to the divergent political, economic, legal, and other circumstances in the two regions, to lead to the expected results.

The insight gained in the project „Transformation of the Central and East European Science Systems" (cf. Mayntz, Schimank, Weingart 1995), that it would be advisable „to become familiar with the socio-cultural and economic concepts and at the same time to acquire a basic understanding of the historical pecularities of the development in each of the individual countries. ... Existing parallels and differences between the transformation processes of the individual nations show that each country goes its own way in the reform of science and research ... " (Filacek, Machleidt 1996 p. 9f.). This reference to the fact that, possibly, respectively different paths have to be found and taken, independent solutions have to be developed, and individual experiences have to be made, is also an allusion to the limits of transferability of „western" models to other systems – that is, of knowledge gained under other economic, political, social, and cultural conditions.

2 Progress

The project's objectives included arriving at results from two different directions: on the one hand, a written „report" on the results of the investigations; on the other, the organization of two workshops – one after the first half, and another at the end of the project's duration.

In preparation for research in the countries concerned – besides organizing discussions with experienced colleagues – we studied the accessible literature, and participated in scientific or technology-political conferences, for example:

Poland – Discussions in Katowice, Krakow, Poznan, Szczecin, and Tychy (Institute for Philosophy of the Silesian University Katowice, Chair for the Didactics of Natural and Engineering Sciences of the Silesian University Katowice, Jagellonian Business School of the Jagellonian University Krakow, the Institute for Philosophy of the University Poznan, the Chair for Philosophy and Business Ethics of the Polytechnic Szczecin, Academy for Management and Social Sciences Tychy, Leon Kozminski Academy of Entrepreneurship and Management Warsaw).

Czech Republic – Discussions in Kladno, Pardubice, Pilsen, and Prague (Ad Vitam Company Kladno, Faculty for Economics and Public Administration of the University Pardubice, Institute for Philosophy of the South Bohemian University Pilsen, Center for Research on Science, Technology, and Society at the Institute for Philosophy of the Czech Academy of Sciences, Prague, Chair for Social Sciences of the Czech Technical University Prague, the Prague Institute of Advanced Studies, the Czech Society for the Environment Prague);

Hungary – Discussions in Budapest (Board of Directors of the Hungarian Academy of Sciences, Institute for Philosophy and the History of Technology of the Technical University Budapest, Institute for Social Conflict Research of the Hungarian Academy of Sciences, OMFB - National Office for Technical Development).

In addition, there were discussions in Moscow (State Committee of the Russian Federation for Environmental Protection, Chair for Social and Cultural Sciences of the State Technical University „Baumann", Institute for Philosophy of the Russian Academy of Sciences, International Independent University for Environmental and Political Sciences), as well as contacts to scientific institutions in Kiew (Ukrainia) and Banska Bystrica (Slovakia).

After the first half of the project's allotted duration, on January 22nd and 23rd,1998, the workshop „Technology Assessment and the Ethics of Science in Central and East European Nations" was held in the European Academy Bad Neuenahr - Ahrweiler; 25 interested parties from various institutions in Poland, the Czech Republic, Hungary, and Germany (among others, the Prague Institute of Advanced Studies PIAS, the Czech Society for the Environment, the Czech and the Hungarian embassies, the Brandenburg Technical University in Cottbus, ITAS Karlsruhe, the Pscherer Educational Instiute Ltd. Lengenfeld, the Coordination Center EG for Scientific Organizations, and the Transfer Center for Appropriate Technology) took part.

The workshop's aims were threefold: *First*, exchange of information about the relevant basic (political, legal, economic, cultural, etc.) conditions, as well as about institutions, projects, and preliminary results in the various countries of the region in question; *second*, promotion of the (until now only hesitantly) initiated expansion of the European TA-network into the Central and East European countries chosen, as well as, *third*: making contacts for bi- or multilateral cooperation.

During the workshop, a first „interim balance" was drawn, and was presented as a formulation of objectives concerning two main themes. In his opening statement, the Director of the European Academy, Prof. Dr. Carl-Friedrich *Gethmann*, characterized the project's status within the scope of this young scientific institution's activities, and then described the Academy's aims and methods. Under the first main theme, „Technology Assessment in Central and East European Nations. The Situation" (Moderation: Prof. Gerhard *Banse*), the situation in the three countries concerned was portrayed. Prof. Dr. Ladislav Tondl (Head of the Center for Research on Science, Technology and Society of the Institute for Philosophy of the Czech Academy of Sciences) and Prof. Dr. Voracek (Faculty for Economics and Public Administration of the University Pardubice) spoke for the Czech Republic. Prof. *Tondl* characterized the situation in the field of TA in the Czech Republic primarily from a scientist's point of view. In this regard, he set the approach for social assessment of technology in the center of his exposition. In his contribution, Prof. *Voracek* concentrated his attention above all on the training of „Systems-Auditors", who are schooled in the University Pardubice, in a program certified by the EU. This curriculum is multi-, or rather, interdisciplinary, and aims at a holistic approach to processes of industrial planning (i. e., by taking political, social, health, economic, technical, and environmental aspects into consideration). Our view of the Czech Republic was rounded off by statements given by Dr. Peter Pechan (President of PIAS), in which he gave information about the difficulties of the practical and institutional anchoring of Technology Assessment in the Czech Republic.

The situation in Hungary was presented by Dr. Tamas Balogh (OMFB - National Office for Technical Development Budapest) and Prof. Dr. Imre Hronszky (Chair of Philosophy and the History of Science of the Technical University Budapest). The subject of Dr. *Balogh's* reflections were the current situation and the aims of research and technology policy. After he had given an account of the staff size, budget, expenditures, distribution of funds, and the promotion of technologies, he gave a short review of the (capricious) history of Technology Assessment in the OMFB. His conclusion was, that, at present, there exist only feeble attempts at TA in the traditional sense of the term, but that instead, tendencies in favor of „technology foresight". are stronger. Prof. *Hronszky* is one of the proponents of TA in Hungary, and is active in the field of politics as well as in that of the sciences. From this standpoint, he gave a survey of the history and present state of TA and similar projects in Hungary, and also delved into lessons learnt from history (e. g., from the project Danube-Dam), into the curriculum, and into disciplines showing an interest in this work.

Prof. Dr. Andrzej Kiepas (Institute for Philosophy of the Silesian University Katowice), as well as Prof. Dr. Jan Such (Institute for Philosophy of the University Poznan) were responsible for the contributions on Poland. While Prof. *Kiepas* concentrated his report primarily on the current state of technical research and TA,

Prof. *Such* emphasized the traditional focus of the Polytechnic Szczecin (formerly Stettin) and of the University of Poznan (formerly Posen) on the Philosophy of Technology.

After each of the „National Reports", there was opportunity for a short discussion. This was eagerly seized by the participants, in order to obtain still more information. It became clear, that, in the countries under discussion, TA-activities are much more strongly a domain of Philosophy than here in Germany, where social sciences are predominant.

For the *second main theme, „Technology Assessment in Central and East European Nations – Opportunities for Cooperation"* (Moderation: Dr. Günter *Lauterbach*, KOWI - Coordination Center EG for Scientific Organizations, Bonn and Brussels), a discussion between the participants from Central and East European nations and representatives of German and West European TA-institutions was planned. To this end, all of the relevant organizations and facilities – from European institutions such as STOA and the German Bundestag, Federal (German) ministries, and German organizations for the promotion of research to TA-research institutions and project sponsors in the Federal Ministry for Education and Research (BMBF) – were invited, in some cases, even personally; several of them accepted the invitation, and replied that they intended to come, but only few actually did take part... (Dick Holdsworth, Head of STOA, at least sent a message of greeting to the workshop participants.) As the first speaker in the treatment of this main theme, Dr. Laszlo *Molnar* (Chair for Philosophy and the History of Science of the Technical University Budapest) reported on „Teaching of and Research on the Ethics of Technology in the Technical University Budapest", and Prof. Dr. Vladimir *Prchlik* (President of the Czech Society for Environment Prague) on environment protection activities in northern Bohemia, in which the Faculty for Environment Protection of the North Bohemian University Usti nad Labem (formerly Aussig on the Elbe) plays a leading role.

In spite of the absence of representatives of many TA-institutions[1], there were in the discussion not only specific questions in the sense of our first objective (exchange of information) but also in accordance with our third aim (making contacts for bi- or multilateral cooperation), and a debate about possible and necessary activities for the future arose. Future topics could be, for example: „Social Technology Assessment", „Technology Assessment and Democracy", „Systems-Auditing", and „The Information Society". Among other agreements, support for the educational program „Systems-Auditing" of the University Pardubice, a program for an exchange of scientists with Poland, the Czech Republic, and Hungary, as well as another workshop in the spring of 1999 were planned. The numerous and varied „informal" discussions held during the pauses were also – as the opinions expressed by the participants show – judged to be very important, reaching as they did from making acquaintance with colleagues to enthusiastic exchanges of information and opinions, from mailing publications to reciprocal invitations. The results of the workshop have since been documented in the European Academy's

---

[1] In the sense of the reasons for the lack of information in Germany about the TA-situation in East Central European countries mentioned at the outset, this lack of presence could be interpreted as an indication for a lack of interest in these countries, which, however, wouldn't contribute to reducing this knowledge gap.

„Graue Reihe" („Grey Series"; cf. Banse 1998a; s. also Banse 1998b, as well as Patz 1999).

At the close of the project, from the 3rd to the 5th of February 1999, the workshop *„From the Information- to the Knowledge Society. Democracy – Participation – Technology Assessment"* took place in Prague, in cooperation with the Center for Research on Science, Technology and Society of the Institute for Philosophy of the Academy of Sciences of the Czech Republik, the results of which are presented in this volume.[2]

3   Results

As results of the project „Technology Assessment and the Ethics of Science in East Central European Nations" („TA - East"), we can list – besides general assessments and the results of both of the workshops, lectures and publications by the project manager during the implementation and follow-up phases – above all, numerous bilateral contacts and cooperations, which have developed in a great variety of forms (agreements of cooperation, initiation of joint projects, invitations to conferences, and to hold lectures or reports, etc.) between the persons and institutions which took part in this project.

Four general conclusions can be drawn:

In the first place – it turned out, above all, that:

* in actual fact, Technology Assessment already was and is being practiced in Poland, as well as in the Czech Republic and Hungary, that is to say, there were „preparatory activities", „competencies", projects and institutionalizations, or rather, attempts at institutionalization.
* This assertion must, however, be qualified by the observation that there is very little practical political, legal, or institutional support for TA, except for expressions of will. Exceptions are the – up to the present, futile – efforts of the Hungarian National Office for Technical Development (OMFB) since 1994, to establish a TA-Bureau as a part of its own organizational structure, and the field of environmental protection, in which many countries have drawn up an Environmental Impact Assessment (EIA), which in some cases is even prescribed by law.

Secondly, these activities, which are often unsystematic or uncoordinated, and dependent on single individuals as well, concern

* „practical" TA as a basis or prerequisite for (political, e. g., investment) decisions, particularly in environmentally sensitive areas (e. g., road building, hydraulic engineering – s. also, merely as an example, the EIA's on the Danube Dam System in northern Hungary, resp. southern Slovakia), increasingly, too, in the question of the „Information Society" (e. g., as to social consequences,

---

[2] During the preparation of this workshop, the organizers published all of the abstracts (cf. Friedrich 1999).

data privacy protection, and demands on education; primarily engineers, natural scientists, and economists are engaged in investigations of this type, occasionally also sociologists – philosophers almost never.
- research on theoretical and methodological aspects of TA, including questions of ethics
- Teaching and training in the field of TA – for engineers and economists, as well as for sociologists and philosophers

In the third place, we may assume that – in the course of the process of transformation in East Central and East European countries in the past years – there have also been changes in the field of Technology Assessment:

- Initially, – often euphoric – attempts to transfer analogously models existing and established in Central and West European countries often predominated. One hoped to be able – alone by imitation or by copying – to produce a situation in which TA could be implemented, and, above all, institutionalized. In this situation, there was an exchange of information, experience, and materials, symposia were organized, and publications appeared – often on the basis of personal contacts.
- The „Sisyphean Toils" which set in after – disappointed – euphoric hopes had to be abandoned, point in another direction, which shows the limits of the "transplantability" of knowledge generated and experience gained elsewhere – i. e., under different economic, technical, political, social, and cultural conditions – to other societies, in the direction of gaining and drawing on one's own experience, developing one's own specific solutions. In this process, „adaptation" – the accomodation of existing models to the respective national circumstances – not just simply copying – played a not inconsiderable role in the development of genuine relationships of cooperation between „East" and „West".
- This odyssey can't yet be brought to an end, because the processes of transformation in the fields of economics, technology, and society, with their conflicting tendencies and regional peculiarities, haven't yet come to a close, the „new" societies haven't consolidated themselves yet. At the same time, new elements are evolving in the „TA-landscape". These are, e. g., interdisciplinary research institutions. New solutions can also be seen in curricula oriented on holistic principles, as well as in the establishment of private foundations as sponsoring institutions for TA-activities.

In the fourth place, we can list the following problems, which are encountered by those attempting to establish TA – or to institutionalize it on a permanent basis – in Central and East European nations:

- Technology Assessment activities are seldom carried out as „concerted", coordinated, converging-action projects (except for some, for instance, in the environmental protection field), but rather as isolated, individual studies. This also means that the results such actions produce are almost never centrally documented and stored, which means: available and accessible; that the individual „actors" in this field often operate without knowledge of one another, to say nothing of a mutual exchange of information.

- TA-activities still find only very little support in the population at large (again with the exception of environment protection – for example, with reference to the North Bohemian region or the Balaton); they are either ignored, held to be unimportant, or – on the contrary – seen as a threat to a rapid rise in the (material) standard of living, or as job-killers. On the other hand, there is almost no public debate about the „pros" and „cons" of concrete technological developments or decisions – not even in the media.

- Often, illusionary notions are encountered in the field of Technology Assessment, whether in the form of trust in „science" or „the" experts in general, in demands for „neutral" or „independent" expert opinions, or in the belief in the state's „omnipotence" and „paternalism" (on the other hand, just the opposite can also be observed, namely, the opinion that TA won't really be institutionalized before enough „pressure from below" has been generated on the basis of a high degree of environmental consciousness).

The information gained to the present can be summarized in the following findings:

First of all: There was and is in fact – albeit with a different history, and in different variants – TA in Central and East European nations: that is, not only has preliminary work been done (which can be built upon), but there is also competence available (which can be drawn upon). On the one hand, it turned out that the standard as well as the methods of TA differ to a certain extent from those in Germany. But on the other hand, new and innovative developments are beginning to appear in this field (for example, industrial organizations and local governments as clients, a more comprehensive view of problems, new conceptual approaches, completely different forms of institutionalization).

Second: it became apparent that an – initially attempted – imitation of practices established in Germany, or in Central and West Europe, because of the different political, economic, legal, cultural, and other circumstances. couldn't lead to the results expected or needed.

Third: it is imperative to note and to document activities in the field of Technology Assessment in East Central and East European countries here in Germany more attentively than has been the case – not only in the sense of „mere" monitoring, but as a basis for recognizing potential links and chances for cooperation[3]. Besides the studies included in the literature mentioned (cf. Banse 1998a), we have already begun to record the relevant information in a TA-database, in order to improve our bibliography with „an eye on the East", and bring it up to date. A prerequisite for this work is – as the project quite clearly shows – that there must be competent and interested partners „on the spot", who can comprehend (and communicate) developments in the field of Technology Assessment in the nation concerned in a bird's-eye-view. Bringing about this condition should be a further objective of German TA-institutions wherever in East Central or East European

---

[3] A study commissioned by the European Academy on the Philosophy of Technology in Russia is certainly to be counted among these opportunities (cf. Gorokhov 2000).

countries this isn't yet or isn't yet sufficiently the case (for example, by means of bi- or multilateral cooperation agreements).

4   Prospects and Visions

The project „TA-East" has, therefore, proven to be of great use, because – on the one hand – a great deal of information about the „landscape" of Technology Assessment in the countries chosen could be gained, and – on the other – the first personal and institutional contacts could be made, in keeping with the objectives formulated by the European Academy Bad Neuenahr-Ahrweiler Ltd. In the course of the project, it turned out that the original goal, which would have been primarily a summary of the available and accessible information, couldn't remain the sole aim, because the opportunities which presented themselves opened new alternatives in the following directions:

- invitations to scientific lectures and conferences
- requests for cooperation and/or participation in committees, and in planning educational programs
- further information about other relevant institutions, and taking up contact with other persons in the three countries chosen, as well as, for example, in Russia, Slovakia, and Ukrainia
- first proposals, resp. concrete steps in planning joint projects on the basis of various funding concepts, including the planning of joint publications.

On this basis, the following perspectives of future research can be formulated:

First: It is necessary not only to keep up the present contacts, but to extend the area „explored" purposively beyond the East Central European countries chosen further eastwards. Important „candidates" would be Russia, Ukrainia, Slovakia, Belorussia, and the Baltic republics (as to Russia and Slovakia, first steps in this direction have already been taken). In this manner, the „TA-network" already existing in West and Central Europe could be extended systematically eastwards.

Second: It is very important to develop contacts further to bi- or multilateral projects. In the course of the project, the following proposals (among others) were made to this purpose: „Technical, Ethical, and Political Aspects of the 'Information Society' ", „Philosophy of Technology in Poland and in Germany, Past and Present – An East-West Comparison", „TA in Eastern Europe – Legal, Institutional, and Practical Problems of Realization in Transferring and Adapting European TA-Education-Programs", „Problems of Sustainable Environmental Policy". These projects could be realized on the basis of various means of funding (e. g., public and private national funds, TEMPUS-program, EU-Framework-program; this would include the planning of joint publications (for example: Principles and Strategies of Environmental Protection).

Third: These considerations have – to the present – taken on concrete form, on the one hand, in agreement on the content of a work-group of the European Academy Bad Neuenahr-Ahrweiler Ltd. on the moral justifiability and cultural manageability of digital signatures (whereby colleagues from Poland and the Czech Republic – among others – should take part in this project), on the other hand, in the form of conceptual considerations regarding a „Four-Nation-Project" (Germany, Great Britain, Poland, the Czech Republic), „Global Information Society and the Application of Information Technologies in Local Initiatives – Social, Economic, and Ethical Aspects" in the Fifth EU-Framework-program.

References

Banse G (1998a) Technikfolgenbeurteilung und Wissenschaftsethik in Ländern Ostmitteleuropas. 2 Teile. Bad Neuenahr-Ahrweiler (Europäische Akademie GmbH) 1998

Banse G (1998b) Technikfolgenbeurteilung in Ländern Mittel- und Osteuropas – erste Ergebnisse eines Projekts. In: TA-Datenbank-Nachrichten, 7 (1998), Nr. 3-4, S. 29-37

BMFT (1991) Technikfolgenabschätzung (TA) in den neuen Bundesländern. Konzepte - Problemfelder - Themen. Hrsg.: VDI-Technologiezentrum Physikalische Technologien im Auftrag des Bundesministers für Forschung und Technologie. Düsseldorf (VDI) 1991

Filacek A, Machleidt P (Eds.) (1996) Transformation of the Central and East European Science Systems. Prague Closing Workshop, December 6-8th 1996. Prag 1996

Friedrich K (1999) Workshop "Von der Informations- zur Wissensgesellschaft. Demokratie – Partizipation – Technikfolgenbeurteilung" 03. bis 05. Februar 1999. Workshopunterlagen. Zusammengestellt von Dr. Käthe Friedrich, Januar 1999

Gorokhov V (2000) Technikphilosophie in Rußland. Bad Neuenahr-Ahrweiler (Europäische Akademie) 2000 (in Druck)

Mayntz R, Schimank U, Weingart P (Hrsg.) (1995) Transformation mittel- und osteuropäischer Wissenschaftssysteme. Länderberichte. Opladen 1995

Patz R (1999) Rezension: Gerhard Banse (Hrsg.): Technikfolgenbeurteilung und Wissenschaftsethik in Ländern Ostmitteleuropas. In: TA-Datenbank-Nachrichten, 8 (1999), Nr. 1, S. 100-102

Schmittel W (1992): Stand und Perspektiven der Technikfolgenabschätzung in den fünf neuen Bundesländern. In: Verbraucherpolitische Hefte Nr. 15/1992, S. 131-149

Technology (1993): Technology in Society. An International Journal, no. 1/1993, pp. 1-160

II The Concept

Science, Democracy and Technological Risks

Jean-Jacques Salomon

Let me first thank the Europäische Akademie, the Czech Academy of Sciences and my old friend Professor Ladislav Tondl for this invitation to take part in your Workshop. Being in Prague is always a deep and immediate awareness and reminding that you are at the heart of Europe, which means the perfect and exceptional blending of tradition, history, art, culture and science. Two years ago I came here for the publication in Czech of my book *Technologický úd_l, The Technological Destiny*, and I concluded the talk I gave on this occasion in the following way, which explains, I guess, why I was invited to open your Workshop and talk about "Science, Democracy and Technological Risks".

We have entered, did I then say, into a new era of the relationship between knowledge and power, an era where the Cartesian rule of methodological doubt needs to be applied to decisions in a context of uncertainty and at the same time of possible irreversibility. I added that the practice of such a doubt no longer can be the monopoly of experts. Facing the power of the technical lobbies in modern societies, there is no other way to limit the damages than to obtain more transparency in the decision-making process, and thus to reinforce the procedures of information, consultation and negotiation which guarantee the democratic functioning of our societies.

It seems to me that what need to be applied in the technological area just reflects the principles of good democratic practices, and from this standpoint, Prague and the whole Czech people have given an admirable example of such rising to speak and even dissenting which led, in spite of so many obstacles, to finding again the path toward democracy. In brief, I concluded that, in facing the positivist approach of the pure technician or the technocrat, it is as important to listen to dissidence, to acknowledge and integrate its legitimacy in the decisions affecting scientific research and technological developments.

You got already the essence of what I wish to say and I could stop here. Actually, speaking in this Academic framework, I will insist on the fact that we are now far from what was the credo of the first Academies in the very beginning of modern scientific research, namely that the pursuit of knowledge is to be dissociated from politics or even values. But let me first refer to the part of *The Technological Destiny* which dealt with the notion of public participation. Behind the word there is more often than not a façade or a ritual, and this is exactly what puts our democratic systems under suspicion. The demand for participation is not limited to scientific and technological questions, but also concerns many other less newsworthy issues. But if the hallmark of participation is to voice specific interests in the name of the general interest, certain technological developments give more reasons for mobilizing opinion: they are met with everywhere, their repercussions concern everybody and often every society, and public opinion is particularly alive to them because their implications transcend the local interest which were the first to be aroused. Moreover, they worry people all the more when they take the form of complicated and even obstruse technical programs because the specialists and decisions-makers concerned are suspected of holding back information.

Whatever the case, the concept of public participation implies that the action of any technostructure can be exposed — and indeed, often, opened up — to inspection by individual citizens or concerned associations. Yet, if participation begins with this right of inspection, it can also stop there without satisfying the revealed need for a more democratic control of the decision-making process. As Sherry A. Arnstein, a political scientists, has shown long ago, there exist various degrees of participation to which decision-makers consent and hence different level of control. Arnstein was in fact dealing with participation in decisions not on scientific and technical matters, but on questions affecting underprivileged American minorities in local government — urban renewal, anti-poverty programs and the so-called Model Cities. But it is easy — and it is revealing that it is so easy — to transpose her provocative analysis to the case of technological institutions (Arnstein,1969).

The fact is that the regulation of technical change can be exploited and manipulated, as much any other case of social regulation. "The idea of citizen participation, says Arnstein without too many illusions, is a little like eating spinach; no one is against it in principle because it is good for you. Participation of the governed in their government is, in theory, the cornerstone of democracy — a revealed idea that is vigorously applauded by virtually everyone. The applause is reduced to polite handclaps, however, when this principle is advocated by the have-not blacks, Mexican-Americans, Puerto Ricans, Indians, Eskimos, and whites. And when the have-nots define participation as redistribution of power, the American consensus on the fundamental principles explodes into many shades of outright racial, ethnic, ideological, and political opposition."

Let us leave aside these material have-nots and concentrate on the outcasts of knowledge — though the distinction is debatable since power depends as much on wealth as on knowledge. What Arnstein brings out, a point as true for the have-nots of the Welfare State as for those of what Don K. Price called the Scientific Estate, is that participation is hollow, frustrating and can lead to revolt when not accompanied by some redistribution of power (Price 1965); it merely allows the technostructure to proclaim that all the "interested parties" have been borne in mind whereas by the end of the process only a selected few have been at best consulted in order to safeguard the status quo. From the mere semblance of participation to the real thing, Arnstein distinguishes eight levels or rungs of a ladder, with each rung standing fore a certain measure of citizen influence over the end-product of the decisions. By grouping them in two's she obtains three broad categories — non-participation, tokenisms and a genuine impact on the decision.

The first category is manipulation and therapy. In the name of participation, an interest is taken in the concerns of ordinary people with the explicit purpose of "educating" them and winning their support. This is really not more than a public relations stunt and can be illustrated by the hastily cobbled information campaigns after an industrial disaster (Seveso, Bhopal) or near-disaster (Three Mile Island). The language of science is used to reassure the local population by imposing the therapy of experts. Just as, according to Arnstein, an effort is made to minister to the "pathology" of the have-nots minorities rather than tackle the racism or inequalities underlying such a "pathology", the arguments of these experts are focused on the "irrationality" of the fears occasioned by the big technologies rather than on the potential threats and real accidents which have engendered these threats.

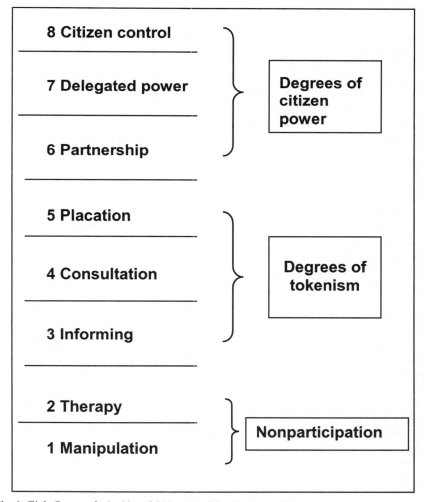

Fig. 1: Eight Rungs of a Ladder of Citizen Participation (Arnstein)

At this level, the citizens are treated like children, handicapped, disabled or diseased persons condemned to beg for scientific assistance. Social control is reduced to a psychological insurance policy which attempts to tune attitudes and values to that "hard facts" school of objectivity to which the technostructure belongs. In order to minister to such irrational fears and superstitious fantasies it turns to the social sciences and even psychoanalysis, which show how this opposition to new technologies is regressive, a rejection of the father, with elements of Oedipian repression and sexual inhibition. I am not joking: this has been a common tone to most nuclear experts and institutions in the 50's and actually later one, as illustrated in a book published in French by a psychoanalyst, *The Atomic Anxiety and Nuclear Plants* (Mendel, Guedeney, 1978). In this book we are given extracts from the 1958 report of a WHO study group on "the questions of mental health raised by the use of atomic energy for peaceful purposes which ends with this gem: "From the viewpoint of Mental Health (*sic* for the capitals), the most satisfying solution for the future of peaceful applications would be the rise of a new generation which has learned to live with a certain degree of ignorance and uncertainty". Moving up from this masquerade or sham we come to the symbolic rite: through informing, people are allowed to hear the arguments of the other side and to state their own views. They are no longer presumed to be "mentally deranged" or children, but are treated as real...adolescents. There is, of course, no guarantee that their fears and suggestions will be taken into consideration, but the participation process has at least shed the guise of a socio-cultural insurance policy to take that of the classroom. At its best, although the information flows in a single direction, it is provided at an already advanced stage of the project: people are no doubt better informed, but the dice have been already cast. In short, this kind of participation is like a formal lecture on the facts of the matter and the conclusions to be drawn in which the audience can do nothing but listen. It is certainly a step forward: better a pittance of information than none at all. Though tokenism may be a mere soporific, its shows the vices of democracy, so to speak, at least making some sort of gesture to its virtues.

At the next rung — consultation — people are invited to give their opinion. Aware of their rights and of their responsibilities, they intervene through opinion polls, attitude surveys or even in public inquiries. They do not, of course, intervene in the decision. As Arnstein rightly observes, they have "participated in participation" and the technostructure can now go ahead, happy with this formal contribution of the "public" to its decision-making. At a still higher level of tokenism, the rules of the game allow the have-nots to formulate their opinion, they are accorded a seat or two on the advisory committees and, provided that their occupants are accepted as recognized spokesmen or women for the interests involved (e.g. have been elected by local associations), the opinions they voice may have a certain impact on the decision even though, for their part, the decision-makers insist on their full right to decide. This is placation: the have-nots feel less excluded and the technostructure better understood.

From one rung of tokenism to the next, the progress is considerable and one cannot underestimate the benefits not only for the public but for the policy-makers too: the insurance policy approach is repudiated, people are treated as adult citizens and above all, in this first attempt of real dialogue, the discussion makes room for fresh considerations, thus setting the issues in a more open, critical and pluralist context. At this level of participation, expert opinion no longer appears exclusively technical but most often than not turns out to be based on presuppositions which reveal the impossibility of drawing a sharp distinction between factual deductions and subjective, if not political, preferences. This is the first attempt to correct the imbalance between the decision-makers and those wanting to have a say in their decisions. By revealing that the arguments of the experts are not solely technical, the participation process abandons the esoteric, univocal and sacred land of "scientific facts" for a more familiar terrain in which facts cannot be separated from values. The decision is taken in public rather than by the public, yet the social regulation of technical change becomes more than the semblance of a "democratic dialogue" since the informing and learning processes are no longer a one way. At this level, there is discussion but still no negotiation.

One has to climb higher, to partnership, delegated power and especially citizen control, before the legitimacy of the decision is genially negotiated between the technostructure and the people. "Partnership" implies that the groups contesting the decision have the financial resources to organize themselves and employ competent scientific, legal or other advisers and then see their views taken into account by the decision-makers. "Delegated power" can be envisaged at the local or regional level when national interests are not at stake. Lord Ashby has cited one of the first cases of such participation process: in 1975, the San Diego Gas and Electricity Company, having to build a thermal power-station in south California, invited the local people to set up a committee to examine possible sites, provided it with financial resources and technical advisers; this committee held its own public hearings and its recommendation was accepted (Ashby, 1980). Since then many other similar cases can be mentioned in Europe as well.

Thus lastly, the top rung of the ladder is "citizen control", which may imply a referendum prepared long in advance by a public campaign, as was the case in Austria or Switzerland in relation to the building of nuclear plants or more recently to genetic engineering in relation, to transgenic plants and new drugs. In sum, if one wants really to go from the mere semblance of participation or from tokenism to real power-sharing, the amount of participation depends on the *decision-makers accepting rules which are no longer laid down by themselves.* In this context, what is at stake is simply the credibility not only of the technical institutions concerned, but also of the whole democratic system.

The trade-off between freedom and security has been the subject of debate ever since the very beginnings of political thinking about democracy. Indeed, in democratic societies security is considered to be a right, and the State is held largely responsible for ensuring it. This is not a prerogative just like any other: the modern State developed in the 17th and 18th centuries on the basis of a "contract" that explicitly linked recognition of the government's legitimacy — and public acceptance of taxation — to a guarantee that the citizens would be protected. The State is permitted a monopoly of legitimized violence on two conditions: that limits are set to the use of such violence, and that the State must take full responsibility for protecting individuals and property. Michel Foucault has defined the ideal government as one that "governs close enough to the dividing line between too much and too little security"(Foucault 1991).

Neither too much nor too little: the ideal balance is *close enough* to the dividing line between one extreme and the other, in a grey area which would be considered quite reasonable if everyone were in agreement. But what happens when there is no consensus? It is even harder to achieve a compromise because we are not dealing here with purely rational attitudes and even worse, with pure objective facts. The contradiction is particularly obvious in democratic societies because they are the ones where demands for greater individual freedom are combined with demands for increased security. Tocqueville noted this tendency already back in the 19th century, and the modern world, with its far greater complexity and heightened awareness of risks to security, still constantly provides examples of the tensions and crises that arise from this contradiction. If security were given absolute priority over all other considerations it would mean either imposing totalitarian rule or complete paralysis: either a police state and a military solution to the problem of risk, or else a rejection of innovation and a timid, unchanging society.

Democracies do not refuse only to face this dilemma; they insist that the rules of the game and the decisions relating to the trade-off between freedom and security are not settled behind closed doors, in the secrecy of what C.P. Snow called "the corridors of power", where decision-makers encounter only technicians and where experts increasingly have the upper hand (Snow 1962). Technocratic rule is usually understood to mean that power is taken away from the representatives of the people. The demand for transparency thus tends to challenge a style of risk management that reduces the complex task of governing human beings to no more than a matter of administering things.

From some angles the demand for transparency seems to be outrageous or even unreal: is it possible to know everything and — what is more — monitor every step in a chain of decision-making? It also means dealing with the risk that decisions will be taken without the people or their representatives having any right to be involved. But lack of transparency means breaking the contract: if governments are to have legitimacy and also if the public is to accept taxation — the two things that underpin the guarantee of security for the governed — then public opinion must be kept informed, i.e. the technical establishment must not withhold information. In areas such as waste disposal or environmental protection, the example of what *not* to do offered by the totalitarian regimes over the last 50 years is so much worse than anything we could have imagined or guessed that one can only agree with Churchill's remark that democracy is the worst possible system ... except for all the rest. I would go so far as to say that in totalitarian regimes the disregard for industrial or technological risks was on a par with their violent treatment of human beings: the two went together.

There are circumstances within democratic societies where decisions are deliberately taken in the dark or in secret, but they are by definition exceptional: war, terrorism, natural disasters. Nevertheless, even in extreme cases, although transparency may be restricted, it is never completely absent. Because it is characteristic of democratic societies that they always demand that decisions be publicized and discussed, even in the most exceptional circumstances — that is perhaps their weakness and also their strength.

The notion of transparency was applied to institutions first in the United States, at the time when authority and the "establishment" were being challenged in the late 1960s. The movement led, among other things, to the "Freedom of Information Act", which allows individuals access to their confidential files held by public institutions. An "open" institution is one that hides nothing. Therein lies the paradox of the demand for transparency: its target is what is supposed to be public but isn't. Any suspicion of the contrary gives rise to complaints that information is being withheld, and to fears of manipulation or even machinations by the institution, which has no intention of sharing its knowledge with outsiders.

In the United States, the way the concept evolved in recent years has as much to do with the Vietnam War as with Watergate: the "credibility gap" with regard to public institutions suggests that — for all that the country is a democracy — its institutions are manipulated and manipulable just as in a society without the same supposed safeguards. In fact, well before the era of major challenges, sociologists like Packard and Mills were criticizing the practices of both public and private institutions that deliberately prevented anyone from seeing precisely what was going on. Galbraith kept up the attack in *The New Industrial State*: a veil is drawn across the truth and the power struggles that take place hidden from the public gaze (Galbraith 1967). There is definitely a link between the increasingly strong concerns about the environment and the demands for greater transparency, which may raise a smile at the Rousseau-like inspiration: the notion of transparency is connected with the notion of original purity, yet surely this is the exact opposite of the industrialized world in which human beings must now live. Lack of transparency in human institutions is thus set against the limpid clarity of nature — but does that really exist?

More seriously, the demand for transparency arises from two phenomena, both connected with the increasingly scientific nature of modern industrial societies: first, technical systems have become so complex that they have to be operated by experts, and secondly, people feel that they are being excluded from any democratic oversight of these activities. The result is the paradox that although these societies have the means of acquiring, processing and sharing information which may give individuals far greater knowledge of the world they live in, this huge growth in the amount of information available is not matched by a similar degree of transparency of society vis-à-vis itself.

On the contrary, the more the media provide instant coverage of world events as they happen and call on "experts" for comments, the more the information creates a vast blur, if not outright faking of "events", as we could all see for ourselves during the Gulf War. The use of experts — turned into technocrats, strategists, prophets or gurus — is sometimes the best way of getting round conflicts and muddling the debate. And while the public may be more eager to be informed, as levels of education rise and the middle class grows, public suspicion is far from being reduced when institutions with specialized knowledge — as we have seen in the case of nuclear power — make efforts to anticipate the public's questions.

No institution likes to acknowledge these areas of vulnerability. To issue warnings of dangers is tantamount to admitting weakness. So how much information can be divulged before it starts to have adverse effects? The mere act of seeking information from various public or private institutions is often unacceptable, and can result in misunderstandings and conflicts. However, the limits to the ideal of transparency are set not only by the producers of knowledge and information, but also by the public's ability to absorb that information. If the object is to elicit a mature opinion, then the appropriate means of information and education must be mobilized more vigorously and more objectively so as to increase public awareness of risk. Instead of avoiding conflict, this educative effort must acknowledge and deal with the deep feelings, interests and tensions that are an integral part of any industrial enterprise. The key aim is not to limit entrepreneurial freedom but to prevent catastrophes — or at least to reduce their frequency and costs to the individual and society.

A report published by the US Department of Energy shows perfectly the link between transparency and the legitimacy of institutions. It is indeed a vivid illustration of how reluctance to be open — or inability to practice transparency, i.e. to admit and recognize vulnerability — can lead a public institution to completely undermine its own credibility with the public for the duration. *Earning Trust and Confidence: Requisites for Managing Radioactive Waste* is the title of this report which was commissioned by George Bush's Energy Secretary, Admiral James D. Watkins, and published in 1993 by his successor in the Clinton Administration, Mrs. Hazel O'Leary. The fact that this report was asked for by a Republican and endorsed by a Democrat is in itself revealing. The title reflects the two aspects of the problem: how to ensure that the public puts its *trust* in the system and feels *confident* about it. This question, put by the Energy Secretary to the chairman of the Task Force, Todd La Porte, professor of political science at Berkeley, touches on problems that are in many ways similar in scale to those in the former Soviet Union (La Porte 1993).

In the United States, the Department of Energy is responsible for managing the nuclear wastes produced by both privately owned nuclear power stations generating electricity and by industries making nuclear weapons, including making good the damage to the environment at sites used for making and testing bombs. Managing these wastes absorbs more than a quarter of the Department's annual budget of over $2 billion, and the US has more than 3700 nuclear waste disposal sites, where the Department is only now starting to measure the levels of contamination. Some of these sites date back to the end of the Second World War and contain vast quantities of radioactive wastes from the post-war nuclear weapons programs. They have often been neglected for years, especially those at Hanford, Savannah River and West Valley.

Furthermore, the decision to create disposal sites deep underground where the waste would be left for ever has naturally encountered vigorous opposition from the states, and even more from the counties, chosen for such sites, for instance at Yucca Mountain. The decision has generated a controversy not only among scientists, but also among the various public agencies involved, with the Department of the Environment at odds with the Department of Energy. Opposition has also come from state governors, members of Congress and pressure groups, especially environmental organizations and Indian tribes whose lands were scheduled as possible sites.

That particular issue aside, the La Porte Report also examined the whole history of the way the Department of Energy had behaved and it was extremely critical, saying that the Department was an excellent example of how a public institution could arouse suspicion, distrust and lack of credibility, not just on the part of environmental groups and local politicians in the states concerned, but even (to its great surprise) on the part of the industries involved in the programs of waste management. The Report notes that this distrust is not "irrational" and cannot be dismissed as another example of nimbyism (the "not in my backyard" syndrome); furthermore it will take a long time to reverse the trend. The Report offers a series of recommendations "that are not just simply choices on a *menu*; (...) rather they represent the panel's recipe for what the Department should do to strengthen public trust and confidence". These recommendations posit a package of conditions defined as necessary though still insufficient which, if implemented, might allow the Department of Energy eventually to become "trustworthy" once more.

The Task Force was made up of social scientists, nuclear experts, industrialists, officials and representatives of voluntary organizations. The inquiry lasted two years, and during this time the Task Force sponsored two seminars, one organized by the National Academy of Sciences and the other by the National Academy of Public Administration, and it heard evidence from many people from both the public and private sectors. I mention this because it shows the Task Force's concern to present their Report as a search for scientific legitimacy in a debate that must have given rise to considerable political tensions. Moreover, the principles stated as underlying the recommendations are presented as "similar to geometric principles":

- "Public trust and confidence is not a luxury. DOE not only has an obligation, but it also has a compelling need to do so.
- Public trust and confidence is not a one-way street. DOE must trust the public before it can expect the public to trust it. By the same token, the public and its representatives must be held to a standard of behavior that is trustworthy.
- Under almost all circumstances currently relevant to DOE's waste management program, it is preferable to make decisions in an open, pluralistic forum than in a closed one that excludes actual or potential stakeholders."

The recommendations run to several pages and call for changes not just in the relations between the Department of Energy and the public, but even more importantly in the Department's approach, methods and presentation with regard to managing nuclear waste — in short, a cultural revolution. The Task Force described "the central contribution that trust and confidence makes to the legitimacy of public organizations within the American system of governance. That contribution derives from a democratic ideology that demands that public institutions operate in a transparent manner, that they adopt processes that not only permit but encourage broad segments of the public to participate, and that no segment finds itself permanently a loser in public controversies".

The Report was of course criticized, like many another produced by the federal departments or any public administration, as being a way of achieving a clear conscience or making up for past errors. On the other hand, some people, including of course members of these departments, and in particular representatives of the nuclear lobby, were attacking its approach as unrealistic and doomed to be ineffectual. Let us put aside the question of how and how much time it will take to reduce such a gigantic amount of nuclear waste and just consider the implications of this lack of transparency. As the Task Force argues with respect to very simple, yet convincing, requirements of efficiency, "in a very concrete sense, it is unlikely that agreements will be reached to manage radioactive wastes absent of a solid foundation of institutional trustworthiness." It is obvious that what refers here to the American situation applies as much to the European one, although the amount of nuclear wastes is certainly less considerable in the Western part at least of Europe. I am ready even to go further than this Report and say that there is a clear link between the suspicion that currently surrounds the functioning of democracies (and indeed the disenchantment with how they work felt by much of the electorate) and the suspicion of institutions whose activities are based on scientific rationality (Salomon, 1992).

In both cases, there is a crisis of legitimacy in Habermas' sense — and each rebounds on the other. It is not a coincidence that Habermas' first work, his criticism of positivism, led him to question the reasoning behind the actions of the modern State: "making politics more scientific" and "making science more political" are both aspects of the same phenomenon whereby technology dominates everything else and is the source of new social conflicts and individual initiatives (Habermas 1975). In all democracies the electorate's decreasing interest in politics is matched by the questioning or discrediting of institutions that are based on scientific rationality.

"Let no-one who is not a mathematician enter here" was, in Plato's world, an invitation to approach philosophy through science. In the contemporary technical system, a sign forbidding entry similar to the ones found on military areas warns off all those who do not know the password or the language of the tribe. The Goliaths of the big-technology networks, the public or private oligopolies of energy production, telecommunications, transport, space and especially defense-related industries are rarely vulnerable to the challenges from the Davids represented by pressure groups or individuals. As the old but still revealing controversy between Jürgen Habermas and Niklas Luhman shows, what is at stake is not just that the political system is becoming increasingly separated from modern society, it is how far individuals are able to make their voices heard in democracies.

For Luhman, transparency is a myth that is all the more dangerous as the gulf widens between open democracy and rationality. There is something old-fashioned in the "old European view" that individuals can influence politics through their demands, when in fact the decision-making process is increasingly independent, administrative and technocratic. And the more that the electorate is urged to take a more active part, the more people are likely to feel frustrated: a decision-making process that pushed them to make explicit choices among the options would simply put them off trying at all. While Luhman highlights the growing independence of the machinery of government from civil society, Habermas is concerned about the increasing dependence of public institutions on powerful, well-organized economic interests.

They agree in emphasizing the limits imposed by the complexity of modern societies on the ability of citizens to make democracy more effective in practice, but whereas Luhman thought that the aim of transparency is both useless and illusory, Habermas considers it to be an essential regulatory principle, "an Idea of Reason" that democracies cannot abandon without ceasing to be democratic. Obviously this approach is the only one that avoids the double trap of technocracy and totalitarian rule. "An Idea of Reason" is an ideal, and if one cannot achieve it, at least one should try to move in that direction: in any case, to merit their name democracies ought never to give up trying to attain that ideal.

This leads me to some concluding remarks about the experts' role in our societies, I mean to underline how little the process of expertise in relation to such technological issues can be dissociated from values judgements. To the technostructure this may appear as undermining further the traditional image of science's neutrality. But this is the price to be paid by an effective democratic dialogue: the scientific advice appears ambiguous as soon as a controversy is extended to the public arena and reveals at the same time the ambivalence of scientists' loyalties. Scientists are trained not to trespass upon the world of judgement and political bickering, and yet they are called upon to take a stance on issues and decisions that cannot be restricted to a scientific assessment.

Take the current case of the threats to the earth's climate: scientists may be alarmed personally at the possibility of catastrophic upheavals in climate change, but as experts they are divided and cannot come down publicly on the long-term trends toward warming because the evidence is not strong enough. Here we are no longer in the territory of science, but in a mixed field of scientific findings and political lobbying. If you read Ross Gelbspan's recent book, *The High Stakes Battle over Earth's Threatened Climate*, you see how much the oppositional view, though marginal, was able within the US Congers to challenge the work of the Intergovernmental Panel on Climate Change (IPCC) to the point that its conclusions appeared faulty and deliberately concocted (Gelbspan 1997). And within the fossil-fuel lobby in the US you find many scientists who may not seem worse from a scientific point of view than those who were members of the IPCC. Indeed, the Policy-makers Summary of the IPCC report leaves out the reservations and presents the bald statements as an agreed conclusion. Moreover, it is striking to see how different the European approach is from the American one. For most of the European countries, as Gene Skolnikoff has shown in a revealing survey, this is now a settled matter and the precautionary principle should apply. But for the US, not only are there some "greenhouse skeptics" in the scientific community, but even when most of the scientists agree on the reality of the threat, they doubt that the temperature rise in the near term can confidently be predicted and thus question the rationale for substantial and costly mitigation measures at this time (Skolnikoff 1997). Uncertainty on one side, possible irreversibility on the other: this is obviously an issue where the interests of the industrial sector plays a role in the policy process as much as political and public opinion in relation to the environment. The latter pushes governments to a strong stance in favor of fuel taxes, the former tends to rely on market mechanisms for achieving cost-effective emissions.

We know, after the Rio de Janeiro, Kyoto and Buenos Aires United Nations conferences, that the matter is far from being settled from a political point of view and that the scientific community is still divided to day. Whatever the results of the future negotiations, this is a perfect example of the ambiguous role that scientists are today doomed to play. And this is really new if one thinks of the predicament of the Charter of the Royal Society, which was given the objective of, on the one hand, "perfecting the knowledge of natural things", and on the other, "not meddling with Metaphysics, Morals and Politics". In our days, scientists cannot avoid being involved as experts in scientific matters that are exposed to the pressure of interests and passions characteristics of the political scene and therefore they are directly immersed in the controversial area of "politics and morals". Such dispute between the pros and cons is no longer a scientific controversy as such, nor the experts' stand in such controversies can be defined as presenting "pure" scientific statements. When experts are brought in, their role does not consist in acting as arbiters, but rather in stressing the facts, consequences and issues at stake which highlight the sources of controversy and conflict. This does not mean that they themselves keep out of the fray: it is one thing to produce scientific knowledge, it is quite another to use it for the purposes of decision-making. As Philippe Roqueplo has shown clearly in connection with the problems related to the greenhouse effect and indeed to environmental problems in general, the credibility of both experts and science is compromised by the different procedures (Roqueplo 1993).

What transforms a scientific statement into an expert opinion is its use in the decision-making process, and from then on the scientist cannot avoid bringing something of him- or herself into the process: his personal and professional background, his sympathies and values, and so on. The objectivity expected of him has to be maintained not only despite pressures from the public or from government but also, no matter how honest he is, despite his own prejudices and commitment. In short, a clear distinction must be made between three separate functions: producing knowledge, offering an expert opinion and making decisions. The scientists' role is to advance the state of knowledge, though with the recognition that knowledge is always under scrutiny. The experts' role is to highlight the issues by providing the relevant information, and to be prepared to enter the realm of controversy. And the policy-makers' role is to weigh up the information and make decisions, accepting the sanction of public scrutiny and the possibility of being removed from office.

It is important not to muddle the different roles or the areas of competence, or — even more — the areas of responsibility. The three elements involve different but overlapping areas of debate, and all three are essential inputs in the process of public decision-making, which is not really a matter of scientific objectivity, but of the clash between strong feelings and interests. In brief, one needs to channel the latent violence by giving it a chance to be transformed into an expression of public will. For this to occur, the decisions must at least be open to negotiation and, for a start, the relevant information must be made available from the very beginning of a project, and not when it is too late for the plans to be changed. But another condition is that the pressure groups have enough money to organize their campaigns and hire competent advisers (scientists, lawyers, etc.); in short, they should not be at a disadvantage compared with the businesses, public or private, whose decisions they oppose, and this may mean the businesses concerned having to contribute to funding counter-investigations.

Let me now conclude: In this case, it is never enough for individuals to have the right simply to oversee the actions of a government department or to have access to information held by that department. More is required for real transparency: something has to be done to correct the inevitable imbalance between the strength of those taking the decisions and the concerns of the individuals or lobbies trying to influence those decisions. In short, it is a matter of renegotiating and redistributing power. Transparency rises above mythology and the easy jibes directed against it when it is clearly understood that the regulation of technologies cannot be left entirely to technicians, and that experts should not have the final say in a debate where the consequences are never purely technical.

References

Arnstein SR (July 1969) A Ladder of Public Participation. The Journal of American Institute of Planners, pp 216-224

Ashby Lord (February 1980), Nature. in his review of Guild Nichols's report on technology and public participation published by OECD in 1979, Technology on Trial, n°283

Foucault M (1991) quoted in J. Theys J., "Postface", Conquête de la sécurité, gestion des risques. (C Douriens, J-P Gaillard et alii edit.), L'Harmattan, Paris

Galbraith JK (1967) The New Industrial Stat., Houghton Mifflin, Boston

Gelbspan R (1997) The Heat is On: The High Stakes Battle over the Earth's Threatened Climate. Addison-Wesley

Habermas J (1975) Legitimation Crisis. Beacon Press, Boston; and his dialogue with N Luhman (1971), Theorie des Gesellschaft oder Sozialtechnologie?. Suhrkamp, Frankfurt

La Porte T (1993) Earning Trust and Confidence: Requisites for Managing Radioactive Waste. Final Report of the Secretary of Energy's Advisory Board, Task Force on Radioactive Waste Management, US GPO, Washington

Mendel G, Guedeney C (1978) L'Angoisse atomique et les centrales nucléaire., Payot, Paris

Roqueplo P (1993) Climats sous surveillance — Limites et conditions de l'expertise scientifique. Economica, Paris. See also the review of this book by A-J Guérin (November 1993) in the journal of the Ecole Polytechnique, La jaune et la rouge

Salomon J-J (1992) Le destin technologique. Balland, Paris (re-issued by Gallimard/Folio, 1994); in Czech, Technologický úd_l (1997), Filosofia, Prague

Skolnikoff EB (August 1997) Same Science, Differing Policies: The Saga of Global Climate Change. MIT Joint Program on Science and Policy of Global Change, Report n°22

Snow CP (1962) Science and Government. Mentor Books, New York

Concepts of Information Society and the Social Function of Information

Gotthard Bechmann

1 Introduction

At no point in time was society able to forecast radical structural change or even to observe such a change comprehensively. The invention of writing, the alphabet or the printing press were hardly noticed and their impact on the reconstruction was not assessed anywhere near correctly. At any rate, contemporaries were unable to diagnose the significance of the event concerned correctly or to forecast the implications of a structural revolution of society as a whole.

It is not simply a matter of multiplication of knowledge. There are structural reasons for the limitation of our potential.

This applies particularly to the subject of the information society. Here, we are dealing with an epochal turning point - or so it is claimed. And changes are affecting societal means of communication and techniques of communication, and thus not simply marginal changes, but the changes affecting the very fabric of social structures.

The term information society implies more of a political programme than a theoretical concept. The path into the information society is expected to lead to a globally competitive economy, creating new jobs and deactivating ecological problems. The information society requires a new form of participation in political decision - making in democracy and according to the promoters of information technology, a revolution of working life and everyday life will be launched. If one makes an attempt to penetrate the veil of wishful thinking and politically superficial hyperbole, one may discover behind the label information society a set of completely heterogeneous theoretical assumptions and conflicting trends describing the information society. There is to date nothing approaching a coherent theory of the information society. In the literature, the information society is frequently characterised by the technologies finding use in it and celebrated as a new technological revolution. Thus the emergence of information society is frequently determined by four interconnected technologically far - reaching innovations:

- The transition of all mechanical, electrical and electromechanical elements to electronic systems
- The miniaturisation of all electronic elements
- The digitisation of communication and information

- The explosion in the field of software development.

The technical description of the information society cannot indicate the social conditions and consequences of this development and to this extent is tautological, since the information society is described as the application of information technology.

More informative are theories proceeding from the cultural assumption that the production, distribution and processing of information has developed into a constitutive condition of social systems in the developed industrial states. There are however widely diverging opinions in scientific and political discussion of how this new significance of information is to be assessed from the viewpoint of society.

However, these changes in communication must be introduced through communication. There is no one outside society who could plan or control this.

The system evolves through self-reference. It can only be steered and controlled by parts of the same system and thus by itself. The observation and description, planning and control of this societal upheaval assumes the society not only as the object but also as the subject of its own activities.

Precisely this is probably the decisive reason why the description and analysis of change in communication structures produce the most varying theories.

Departure for the information society is taking place on the many roads that, as is well known, all lead to Rome. Another typical aspect of the situation is probably that scientific analysis and political proclamation are in many cases inseparably entwined. Precisely because the future is unknown, there is a need for orientation and science seems to have occupied Cassandra's place in this - with all of the problems linked with erring and admonishing.

The development of the information society is at the same time an evolutionary event and political will. Thus it is little wonder that Kubicek was able to identify three approaches during the past twenty years in the Federal Republic of Germany to find its way into the information society (Kubicek 1996).

In 1974, the then ruling social-liberal Federal Government set up a commission on the development of technical communications which was to determine which telecommunications services were "economically sensible" and "socially desirable" (K+K 1976). The main focus was thus on communication of data and two-way cable television, and thus basically on a technological route into the brave new world.

In the year 1984, the conservative-liberal Federal Government started a programme to coordinate the support programmes of the various ministries (BMFT 1984). The Federal German Postal Service was to invest c. 500 billions of DEM over a period of 20 years for the upgrading of the analogue telephone network into a cable switched broad band network with a glass fibre connection for each private household in Germany (BPM 1984). At the same time, the German Federal Postal Service began to construct a cable television distribution network based on coaxial cables. The conflict between glass fibre and copper cables was born (Kubicek 1985). This time again, it was mainly a technological initiative.

In the meantime, the technological supply side has multiplied several times so that the third attempt to enter the information society took place under the umbrella heading of "multimedia". The Bangemann-Report of 1995 was in a sense the bugle call heralding the collapse of the wall of Jericho. But once again, interest in new technologies and their implementation were in the foreground.

While we are unable to observe and describe future society, we might yet be able to observe what kind of structural change is taking place. We might be unable to position the event between before and after, but we are at least in the position to recognise in which respect the fundamental boundaries of existing societal structures are changing. This is precisely the goal being pursued by the theories of the information society. To this extent, they have a common underlying problem: Tackling the issue of social change due to changed communication and interaction opportunities.

The following discussion of the theories of the information society attempts to provide a brief historical sketch of the theoretical investigations, at the same time aiming to focus systematically on those aspects of social development dealt with from the perspective of the information society. Any theoretical communication is at the same time also a part of social reality and the description of a change to which it itself belongs.

2   Concepts of Information Society

2.1
The Information Society as Information Economy

With the advent of information technologies in the broadest sense, the economic sector of society has become the special focus of economically oriented analyses: The economics of information was created. This direction was principally developed in Japan and the USA. Strictly speaking, information economics embrace two approaches: The "industry approach" and the "occupation approach".

The "industry approach" is based on a three - sector model - similar to that used by Fourastié -here also a shift in relevance between the sectors is being observed. Thus, during the last century, one may notice a shift of emphasis and function from agriculture to industry and from industry to the information sector.

In order to operationalise this development, two indicators have been developed, mainly in Japan: the information rate and the information index.

The information rate mirrors the relationship between total spending of a household and its spending on information.

In contrast, the information index measures the consumption of information. Examples for indicators are the number of telephone calls per person and unit of time, the number of cables per 100 persons, number of television sets, telephones, radios and computers per household, the volume of the service sector, the number of students in the population etc. On the basis of these calculations the "post industrial information society" is defined more or less arbitrarily.

The "occupation approach" of information economics has as its starting point the analysis of the occupational structure of society. It was developed mainly by Fritz Machlup and refined and differentiated by Porat.

Machlup assumed that another sector was emerging in addition to the three economic sectors agriculture, industry and services, and termed this the information sector. This sector consists of various areas such as education, research, the media, information services etc. The constitutive trait of the information sector are occupations and groups of occupations doing information work. By this Machlup (1962) means the production, processing and distribution of information.

Information workers are subdivided into information producers and information users. Porat (1976) has differentiated this classification even further, so that in the end almost all occupations were again to be discovered in the information sector.

By means of this classification, it seemed possible to "measure" information workers and the degree of informatisation of various societies by comparing their occupational structures.

Information economics with its two approaches is confronted with two debacles. The underlying concept of information is unclear and the distinction between information work and other kinds of work remains unsatisfactory.

Thus Drucker defines information as data which are processed with the computer. The Japanese investigation (RITE) defines information as rows of symbols which achieve meaning through social actions and Porat regards information as data which are organised and communicated. The category "information work" also remains controversial. While there have been various attempts to define the term "information work", these approaches have restricted themselves to defining the conversion of input into output of data as information work. The determining criterion which is supposed to distinguish the information society, namely information itself, remains unsolved in this theory.

2.2
Information Society as a Post - Industrial Society

In contrast with the economic and deterministic "information economy" approach, the sociologist Daniel Bell has developed the analytical concept of a "post-industrial" society, in which social change is modelled as a multi-dimensional process. In this process, the various axes of organisations within industrial society (sectors of industry, groups of occupations, the foundations of technology, guiding societal principles) are changed so fundamentally that one may speak of a post-industrial society. Post-industrial society is based on a growing importance of the information sector in comparison with the production of goods, on production primarily dependent on information (instead of raw materials and energy) and also on a changed "axial principle" (guiding orientation of society).

While it is customary in industrial societies to produce goods primarily with machines and private property is regarded as the axial principle, in post - industrial societies concern is mainly on the production and exploitation of information and knowledge. This is made possible by substituting intuitive judgements by algorithms in the shape of intellectual technology based on computer - supported information processing, and a comprehensive scientification of all areas of life (Weingart 1983).

By this means, theoretical knowledge is becoming a guiding principle of social organisations and the post-industrial society is developing into a knowledge or information society (cf. Bell 1989, p112ff. and 353; as a criticism Stehr 1994).

For Bell the drivers of the transition to the information society are primarily technological innovations (especially in microelectronics), and the exponential growth and differentiation of knowledge. The quantitative and qualitative growth of the stock of knowledge and techno-economic growth of the information sector have a mutual relationship in this process, although they are also induced by further dimensions of social change, for example the growing importance of the science system (cf. Bell 1989, p179ff.).

In the political context of the seventies, when societal planning based on scientific methods was still regarded as acceptable, it is easy to see why Bell assumed that scientifically informed decisions would gain importance in relation to other means of societal control (at that time, Dermont 1974 took a more sceptical view).

A glance back at the introduction and diffusion of telematics (telecommunications and computer science) reveals that information, regarded widely as a basic trend in society, adheres very closely to the immanent rules of systems organised according to principles of market economy. It is not so much theoretical knowledge produced by such things as technology assessment as primarily maximisation of profits, competition, technological pressure to adapt and the state as an aggregation of individual economic interests that determine the dynamics and modalities of the information society (cf, Werle 1990, Kubicek and Berger 1990, p45ff.).

To this extent, information society is to be conceptualised not so much as a post-industrial society, but as an informatised industrial society determined by market economy.

2.3
The Information Society as a Knowledge Society

The theory of the knowledge society links with Bell's "post-industrial society" but criticises that this elevates knowledge to a new axial principle of new post - modern society, while not itself having an adequate concept of the meaning of societal knowledge, since it only inquires functionistically about the impact of the growth of importance of theoretical knowledge in modern society, but not about the socially determined processes of distribution and reproduction of knowledge and its reception in society (Stehr 1994, p249).

Generally, a society is termed as an information society if its main conditions for reproduction depend on scientifically created knowledge. Increasingly, scientific knowledge is becoming the single source for generally acknowledged knowledge. This is not to say that scientific knowledge is the only resource of society for knowledge. Apart from science there are still common wisdom, religious knowledge, poetic intuition etc.

It is, however, decisive for the transition to the knowledge society that knowledge implies that participation in the cultural resources of society, and at the same time through scientific knowledge the material conquest of nature, is transformed into a scientifically controlled process. In this, important component systems are becoming dependent on scientific knowledge.

However, this knowledge is a strange product. It is a public good and in principle accessible to all members of society to the extent that its specific conditions of production stick to the rules. It is always only hypothetical and thereby can easily be challenged. At the same time it is permanently creating new opportunities for action and, in doing so, is the basis and the motor of progressive change. In the meantime it has been recognised that science not only creates knowledge, but also simultaneously creates lack of knowledge (Japp 1997). The debate on the risk society, and also the disturbing ongoing ecological discourse during the past 20 years in advanced industrial states, has made it obvious that lack of knowledge is a source of societal argument (Ladeur 1995, Luhmann 1992).

If we assume a threefold division of societal knowledge, namely interpretative knowledge, production knowledge and action knowledge, one may observe that the production of knowledge for these three types is controlled by science in modern societies.

The scientification of society is combined with increasing reflexivity. This refers to the fact that societal structures and processes in knowledge societies are almost exclusively accounted for by decisions that knowledge is continually fed back into actions (Giddens 1990:). All meta - social links are broken and even religion is interpreted as social action, i.e. figuratively socialised. Thus, the information society lacks any kind of non - societal transcendental frame of reference.

On the one hand, scientific knowledge is transferred in the shape of natural science knowledge into technologies and rationalises the relationship of society with nature, on the other hand social science knowledge is transformed in action and decision - making contexts, thus rationalising the meaning and interpretation system (culture) of society. The consequence is that the societal component systems are increasingly organised along the principles of knowledge (scientification), but precisely this increasing scientification does not create certainty. Instead, new dangers, uncertainties and intransparencies emerge, creating the need for more and new knowledge.

A knowledge society is a society which, because of an inward motor, creates both ever more knowledge and also a need for more knowledge by means of the same process. On the basis of this structural description of the knowledge society, Stehr distinguishes between knowledge and information. Knowledge is generally the capacity for actions while information is knowledge processed for the purpose in hand (Stehr 1994, p241f.).

The extent to which this distinction is valid will be discussed further below. Possibly one could already be more precise by stating that knowledge means the structural aspects and information the process aspect of communication, since communication cannot be conceived in any society without concurrent knowledge and societies only differ with respect to their ways and means of organising these two aspects.

2.4
The Information Society as the End of the Mass Production Society

The forceful development of new information and communication technologies and their massive application is linked closely in a tradition emerging from industrial sociology with the change in societal work, its forms of organisation and the related restructuring of the economy during the last quarter of the twentieth century. In this connection, one speaks of the transition of the Fordist-Keynesian era to the post-Fordist era.

Very roughly, the label Fordist-Keynsian era refers to the period from 1940 to 1980, which was characterised by the expansion of mass production and mass consumption.

Mass production implies the standardised production of goods in gigantic production plants such as the Ford factory, chemical plants or other large-scale industrial installations. This kind of production is characterised and symbolised by the conveyor belt. The dominant groups of occupations are industrial workers, who mainly do manual work in whatever form.

Against this background only the disruptions launched by information and communication technologies are observed. The collection, evaluation, processing, transposition and distribution of data - these are the occupations which increasingly play the main role in human labour and which are edging the immediate production work related to manufacturing out of its previously dominant position.

An information economy is emerging which does not coincide with the general term of service economy, but has its specific trait in the process of the creation and processing of information. The organisation of enterprises is changing, in a sense mirroring the change of work.

On the one hand, we are witnessing the emergence of the transnational enterprise which operates free of any national orientation, internationally and independently of location. Outsourcing of areas of production to different areas of the world indicates a partial dissolution of classical firm structures. Under the pressure of customer demand and international competition, "individualised mass production" is emerging. The downsizing and slimming of the enterprise (downsizing, lean production), the levelling of hierarchies and the holistic use of labour in group work, permanent learning by staffs, concentration on competencies, and outsourcing of all other functions to subcontractors are the prominent features of this means of production. At the end of this process we have the enterprise which is determined by information economy and which will provide the central institutional frame for the future of work - the "virtual enterprise".

Forms of work are flexibilised with the help and not on account of information technologies and beside the classical normal worker relationship, a broad spectrum of forms of employment are emerging: tele-work, new self - employment, part - time workers are the keywords indicating the change.

The decisive point is that the new information and communication technologies are the means and the medium for these processes of change.

Means to the extent that they are applied to exploit existing possibilities for rationalisation to make work more productive.

Medium to the extent that they simultaneously create new demand and new forms and content of work. Talking about information work deliberately gives expression to this ambivalent relationship. In the past, work was also always linked with knowledge and information, with the difference that today the aspect of information collection, processing and distribution has become an important element of the process of production at all levels of the organisation of work, manufacturing of goods and their distribution.

2.5
Dimensions of Information Society

We have presented a number of theoretical concepts, all of which analyse the emerging information economy or society, however, each from a very different perspective. Although each approach sheds some light on the significance of information and information technology in the current transformation process, no single one is capable of encompassing all the different aspects of change. It is, of course, difficult to separate the various aspects from each other clearly; they overlap to some extent and some convergence can also be seen. However, whatever attempts are made to develop a comprehensive approach, we are still far away from any general theory of information economy or society.

Do the various approaches to information society have anything in common? Obviously all proponents of information or knowledge society agree that the emerging new society will bring about change at the core of society as it initiates a new mode of production. It changes the very source of wealth-creation and the governing factors in production. Labour and capital, the central variables of the industrial society, are replaced by information and knowledge as the main variables in the new emerging society. But, as Castells (1997) argues, what may be even more important is the establishment of a new mechanism of immediately applying new information and knowledge in the production process. Here the focus is on learning.

Despite the widespread agreement that the emerging information society will cause major social changes, there is still a debate on whether we can speak of information society as a totally new society. This argument is based on a more general tendency to look at historical transformation in terms of the shift from one 'phase' or 'stage' of development to another. Often the socio-economic transformation is seen as a movement from bad times to a more promising and challenging future (Miles and Robins 1994, p15).

This linear perspective, of course, is very simplistic; it overestimates change and does not recognise continuities over time. Certain new tendencies in the present are seen as key elements of the future, while countervailing and divergent processes are not taken into account: "Rather than seeing economic and social change in this linear way, a more appropriate metaphor might be one of cumulative developments in which the new developments form as 'layers' across the old, with new and old always co-existing" (Miles and Robins 1994, p17). From this we can conclude that the increased use of information and knowledge in production does not necessarily include a fundamental change of the character of our existing market economy. It seems to be more appropriate to talk about a new information and knowledge-based industrial and service production.

The role of ICTs in the current transformation process is also highly controversial. In a research area which is mainly concerned with new technologies there is always a danger of falling into a perspective of technological determinism. But while some scientists and politicians still argue that ICT is the driving force behind the emerging information society, others are trying to avoid such argumentation. However, though the social is taken into account, this is often done in a more functionalist way: The argument then is that techno-economic change brings about new social structures and institutions.

Some researchers, more conscious of the temptation of technological determinism, point to global competition, changing consumer demands or deregulation as important factors that cause the transformation process. Still, modern ICTs are given a major role in the transformation process. They can, it is argued, support the development of new social structures, given a push by other drivers of change.

3   Information Technology and the Communication Process

As if they were enchanted by a long humanistic tradition, following the invention of the computer, people first asked what this implied for human consciousness. With the distinction human/technology it is possible to observe how human facilities can be replaced by technology, in our case the computer. The whole direction of "artificial intelligence" research has been inspired and driven by this issue.

Fascinated by the performance of the computer in the reproduction of human forms of consciousness, people have tended to underexpose the societal function. If one does not start from the distinction human/computer, instead attempting to formulate the significance of information technology for societal communication, other impacts draw our attention. However, for this purpose we require a more precise determination of what is to be meant by communication.

Without wishing to entangle myself in the long controversies on the concept of communication, I would like to adopt the systems theoretical concept, which I regard as the best thought out approach following an examination of several definitions of communication. Luhmann (1984, p193-212) deliberately denies the equivalence of communication and action. For him, communication is a social process which is closed in the self-referential sense. Communication takes place due to a synthesis of three different selections.

Like life and consciousness, communication is an emergent reality, a fact sui generis. It takes place through a synthesis of three different selections – namely selection of an information, selection of the message conveyed by this information, and selective understanding (comprehension) or misunderstanding of this message and its information.

None of these components can take place separately. Only together do they create communication. Only together – that means when their selectivity can be brought into a state of congruence. Communication can only take place if there is first perceived to be a difference between message and information. This is the difference from the mere perception of the behaviour of others. In understanding, communication registers a difference between the informative value of its content and the reasons for transmitting the content. In doing this, it can emphasise either side, that is, it can give more attention to the information itself, or to the expressive behaviour. However, it is completely dependent on experiencing <u>both</u> as selection, <u>in doing so</u> making a distinction. In other words, it must be assumed that the information cannot be understood as it is and that to convey its message, a special decision is required. This naturally also applies if the messenger is conveying a message about himself. If and to the extent that this separation of selection does not take place, there is simple perception.

Not only information and conveying the message, but also comprehension itself is a selection. Comprehension is never a simple duplication of the message in a different consciousness, it is rather the continuation condition in the communication system that is the condition for the autopoeisis of the social system. Whatever opinion the participants might have of this in their own self-referential closed consciousness: the communication system reproduces its own understanding or misunderstanding, and for this purpose creates processes of self-observation and self-control.

It is possible to communicate an understanding, misunderstanding and not understanding – however only under the highly specific conditions of the autopoiesis of the communication system and not simply as those involved might wish. The message "You don't understand me" thus remains ambivalent, at the same time conveying this ambivalence. On the one hand it says "You are not prepared to accept what I wish to tell you" and attempts to provoke admission of this fact. On the other hand, it is the conveyance of the information that communication cannot be continued under this condition of not understanding. And third, it is the continuance of communication.

What is new about this concept of communication? There are four specific aspects of this concept of communication worth noting:

- A novel aspect is that the components of communication information, message and understanding only cause the emergence of communication in a synthesis.
- There is a similar distinction in the works of Karl Bühler (1965) from the viewpoint of different functions of speech communication. Americans like Austin (1989) and Searle (1994) have reinforced this into a theory of types of acts or speech acts. Jürgen Habermas (1981) has in turn linked this with a typology of validity claims implied in communication. However, all of this still regards the communication process as the successful or unsuccessful transmission of news, information or assumptions of understanding.
- A novel aspect is that the three components information, message and understanding, are not interpreted as functions, actions or horizons for validity claims, although there is no need to dispute that these are possible means for their use. They are rather different selections, whose selectivity and whose range of selection is first determined by communication. There is no information outside communication, there is no message outside communication, there is no understanding outside communication – and this is not in the causal sense, where information is the cause of the message and the message should be the cause of understanding, but in the circular sense of mutual conditions.

- A novel aspect is that communication as a completely closed system creates the components of which it consists. In this sense, the communication system is an autopoietic system which produces and reproduces everything which functions for the system as a unit through the system. Only communication can influence communication.
- A novel aspect is that communication is seen as having no purpose. Communication has no purpose, no immanent entelechy. It takes place or not – that is all that can be said about it. To the extent that autopoiesis is functioning, it is of course, possible for purpose-directed episodes to form within communication systems. But these purposes are mere episodes without being a purpose of the system. Any other opinion would need to explain why the system continued to exist after achieving its goals, or one would have to say, without being exactly original, that death is the purpose of life.

4   Data, Knowledge, Information

As the former discussion has demonstrated, the concepts of information and knowledge society are not very clearly distinguished from each other. Therefore it becomes necessary to acquire a better understanding of what is meant by 'information' compared to 'data' or 'knowledge'. In the following we will try to differentiate more clearly between the three concepts. Data are, according to Bohn, "what comes directly from sensors, reporting on the measured level of some variables" (1994, p61). They are "events or entities represented in some symbolic form and capable of being processed" (Earl 1996, p38).

Only when data have been organised do they become information. We can define information as "data that have been organised or given structure - that is, placed in context - and thus endowed with meaning" (Glazer 1991). Information therefore includes a process of manipulating, re-presenting and interpreting data. The aim of developing information is to reduce uncertainty or ignorance, give surprises or insights and allow decision-making (Earl 1996, p13).

Information and knowledge are distinct insofar as it takes knowledge to produce valid and useful material and to be able to interpret this material (Miles et al. 1995, p16). While information, according to Bohn, tells the current or past status of some parts of the production system, "knowledge allows the making of predictions, causal associations, or prescriptive decisions about what to do" (1995, p62). It is important to understand knowledge as an active process; it involves the ability to interpret data. It is not just the content of a data base (Miles et al. 1995, p16). The tacit aspect of knowledge is "linked to the fact that production technology is about doing things, not only about knowing things in the form of an abstract (scientific) principle. Operating know-how is something very different and is much less transmissible" (Cohendet and Llerena 1997).

Taking up the idea that knowledge exists in different forms, Earl (1996) suggests differentiating between three aspects of knowledge: science (which may include accepted law, theory, and procedure), judgement (which may include policy rules, probabilistic parameters, and heuristics), and experience (which is no more than transactional, historical, and observational data to be subjected to scientific analysis or judgement preference and also to be a base for building new science and judgement). The author speaks about a hierarchy of knowledge: experience can be characterised as potential knowledge, judgement as workable knowledge and science as accepted knowledge. This means that each ascending level represents an increasing amount of structure, certainty, and validation.

Earl associates the three different levels of knowledge with different categories of learning. "Experience requires action and memory, judgement requires analysis and sensing, whilst science requires formulation and consensus (Earl 1996, p41). He also postulates the hierarchy of knowledge to be synonymous with the distinctions between data, information, and knowledge. "The lowest level is the equivalent of transaction data (and transaction processing systems). The middle level is the equivalent of information in the classical sense of reducing uncertainty to make decisions (and thus equivalent also to decision support systems). The highest level is knowledge where the user is constrained only by its availability or the intellect to exploit it (and thus approximate to the classical expert system or what some call intelligent knowledge-based systems)" (Earl 1996).

Against the background of this concept of communication, one can state more clearly how the concept of information can be determined. More precise definitions of the concept of information often follow the description by Norbert Wiener, "Information is information, not material or energy" (Wiener 1948, p155).

Information can be distinguished from material and energy, although it is usually subsequently treated like both. Information can be measured, transported, saved, stored, sold and bought. Simultaneously, it is also seen that information has a surprise value and that it changes the systems in its state, thus in a sense having impact on the cognitive structure of the information processing system.

It is common practice to combine both in the concept of information, surprising selections and the metaphor of transport, even if both might contradict each other. Jürgen Mittelstraß makes a distinction between knowledge and information: "We speak of information as if it were already the whole of knowledge and, in doing so, overlook that information is only a special form of knowledge, namely the way and means by which knowledge is transported.... In the place of an own competence for the formation of knowledge, there are processing competence and the trust that information is "correct". – One has to believe information if one cannot check its knowledge, the knowledge becoming visible through information" (1993, p24).

If one comprehends information as a selection of communication contents from a repertoire of possibilities, one may recognise that information is not a stable, transportable, workable entity, but an event which loses its character as information through up-dating.

Thus one must distinguish information and transferable knowledge. Interest in information is driven by the charm of surprise. It is the difference between that which could be the case and that which has just taken place or has been conveyed. As a difference it has neither dimensions within which it could vary, nor a location in which it could be found. It is an operation within the communication process. One can merely denote the system which is concerned with it. At the same time, it causes changes within the system. It is thus a qualified difference, or as Bateson (1981, p582) states, "Information is a difference which makes a difference."

The gains of certainty which can be achieved through information are thus always linked with surprises and present certainty as contingent, thus as possible as a difference. Furthermore, information can only surprise a single time. If it is known, it retains its meaning, but no longer informs, only creating redundancy. Its meaning can be repeated, but not its character as a surprise.

Information is a deeply ambivalent matter. It is at the same time an event and a difference. It is a double-sided concept: it helps as it disturbs. In a sense, it contains its own counterpart. From one minute to the next it continues to reproduce knowledge and non-knowledge. As an information it produces continuance opportunities, but on the other side it renews the background knowledge that there are other possibilities. Information must not be correct, just plausible. It must enable the crystallisation of sense and thus permit the continuation of operations and the transfer of the ambivalence of knowing and not knowing to the next situation. To this extent, the information society is chronically uninformed.

5   Conclusion

What is historically new is not to be recognised in the shape of technical and institutional accomplishments, such as mass media, computers, industrial technology, but more in the form of formally ambivalent concepts for the description of modern society. Paradoxically, the information society then appears as structurally uninformed, since each communication also communicates the opposite side of the selection.

Viewed from the societal aspect, information society implies that communication is structurally linked with technology (technical networks), but that information itself is produced outside the technical networks. This leads to increasing susceptibility of communication to disturbances, but also to an immense explosion of communication opportunities.

The manner in which societal communication has been changed in its forms by the use of the computer is more important and more interesting. Computer-mediated knowledge demonstrates the existence of universally operating connectionist networks for the collection, treatment and re-publication of data, which operate specified for subjects, but not temporarily and spatially restricted.

A new relationship between surface and depth is emerging. The surface is the monitor with very restricted demands on human senses, while depth are the networks and machines themselves, providing access to innumerable data, which are restructured and expanded from one moment to the next, transforming them to information.

The creation of a virtual reality denotes precisely this process, by which the opportunity exists to create in principle perpetually new information and worlds from the stored data by combining surface and depth.

The information society is the consequence of such experience that the whole world may be communicated. The phenomenology of being is replaced by the phenomenology of communication. There is the danger that the difference between information and conveying a message will increasingly disappear and be lifted.

How does the selectivity of communications rearrange itself under these conditions?

On the one hand, technology imposes one-sidedness of communication. One no longer selects **during** communication but **for** communication. Communication then takes place as though in a hyper-cycle of mutually dependent selections, but if and to the extent that it takes place, it no longer has the ability to correct itself (television).

Computer-mediated communication enables the input of data and the retrieval of information, but it separates these to an extent that there is no longer any identity in the process itself. In connection with communication, this means that the unity of conveying the message and understanding is abandoned.

The authority of sources of communication with all of its required socio-structural precautions (stratification, reputation) is no longer essential, or even nullified by technology and is replaced by the anonymity of the source. That is, the unity of conveying the message and understanding is abandoned.

The possibility to recognise the goal of a message is also lost. What takes place is an absorption of uncertainty, which to a limited extent controls itself.

References

Austin JL (1989) Zur Theorie der Sprechakte, 2.ed. Stuttgart Reclam

Bateson G (1981) Ökologie des Geistes: Anthropologische, psychologische, biologische und epistemologische Perspektiven, Frankfurt/M Suhrkamp

Bangemann M (1995) Europas Weg in die Informationsgesellschaft. In: Informatik Spektrum 1, pp 1-3

Bell D (1976) The Coming of Post-industrial Society: A Venture in Social Forecasting, New York: Basic Books

Bell D (1989) Die dritte technologische Revolution und ihre möglichen sozioökonomischen Konsequenzen. Merkur 444, pp 28-47

Bohn RE (1994) Measuring and Managing Technological Knowledge. In: Sloan Management Review 12, pp 67-85

Bühler (1965) Sprachtheorie: Die Darstellungsfunktion der Sprache. 2.Aufl. Stuttgart Klett

Castells M (1997) The Rise of the Network Society. Maldon Mass./Oxford: Blackwell Publishers

Cohendet P, Llerena P (1997) Learning, Technical Change and Public Policy: How to Create and Exploit Diversity. In: Edquist C (ed.) Systems of Innovation, Technologies, Institutions and Organizations. London: Pinter

Earl M (1996) Knowledge Strategies: Propositions From Two Contrasting Industries. In: Earl M (ed) Information Management. The Organizational Dimension, Oxford: Oxford University Press

Enquete Kommission (1998) Deutschlands Weg in die Informationsgesellschaft. Zukunft der Medien in Wirtschaft und Gesellschaft. Deutscher Bundestag (ed.) Bonn: ZV Zeitungs-Verlag Service

Fourastie J (1954) Die große Hoffnung des 20. Jahrhunderts. Köln

Fulk J, deSantis G (1995) Electronic Communication and Changing Organizational Forms. In: Organization Science, Vol. 6, No 4, pp 337-349

Gershuny J (1983) Social Innovation and the Division of Labour. Oxford University Press

Gershuny J, Miles I (1983) The new service economy: The transformation of employment in industrial societies. London: Frances Pinter

Giddens A (1990) The Consequences of Modernity. Standfort, Standfort University Press

Glazer R (1991) Marketing in an Information-Intensive environment: Strategic Implications of Knowledge as an Asset. In: Journal of Marketing, 55

Habermas J (1981) Theorie kommunikativen Handels. 2 vol. Frankfurt/M Suhrkamp
Ito Y (1989) Information Studies Today. In: Schenk M, Donnerstag J (eds.) Medienökonomie. München
Japp KP (1997) Die Beobachtung von Nichtwissen. In: Soziale Systeme 3, pp 289-312
Jessop B (1991) Fordism and Post-Fordism. A Critical Reformulation. Lancaster Regionalism Group, Working Paper 41/1991
Kern H, Schumann M (1984) Ende der Arbeitsteilung? Rationalisierung in der industriellen Produktion: Bestandaufnahmen Trendbestimmung. München
Kern H (1994) Globalisierung und Regionalisierung in industrieller Restrukturierung. In: Krummbein W (ed.): Ökonomische und polistische Netzwerke in der Region. Beiträge aus der internationalen Debatte, Münster
Krippendorff K (1994) Der verschwundene Bote.Metaphern und Modelle der Kommunikation. In: Merten K, Schmidt SJ, Weischenberg S (eds.) Die Wirklichkeit der Medien. Opladen Westdeutscher Verlag pp 79-113
Kubicek H (1985) Die sogenannte Informationsgesellschaft. Neue Informations- und Kommunikationstechniken als Instrument konservativer Gesellschaftsveränderung. In: Altvater E u.a. (eds) Arbeit 2000. Über die Zukunft der Arbeitsgesellschaft. Hamburg VSA-Verlag pp 76-109
Kubicek H (1996) Deutschlands dritter Anlauf in die Informationsgesellschaft. In: Buhlmahn E u.a. (eds.) Informationsgesellschaft - Medien - Demokratie Marburg BdWi-Verlag, pp 241-268
Kubicek H, Berger P (1990) Was bringt uns die Telekommunikation? ISDN - 66 kritische Antworten. Frankfurt am Main/New York
Löffelholz M (1993) Auf dem Weg in die Informationsgesellschaft. Konzepte Tendenzen - Risiken. In: Hessisches Kultusministerium (eds.). Zukunftsdialog zu Lehren und Lernen: Angebote. Wiesbaden
Ladeur KH (1995) Das Umweltrecht der Wissensgesellschaft. Berlin Duncker&Humblot
Luhmann N (1984) Soziale Systeme. Frankfurt/M Suhrkamp
Luhmann N (1992) Beobachtungen der Moderne. Opladen Westdeutscher Verlag
Machlup F (1962) The Production and Distribution of Knowledge in the Uniteted States. Princeton Princeton University Press
Miles I (1988) Information Technology and Information Society: Options for the Future. London ESRC
Miles I, Robins K (1994) Making Sense of Information. In: Robbins K (ed.) Understanding Information Business, Technology and Geography. London/New York: Belhaven Press pp 123-155
Miles I et al (1995) Knowledge-Intensive Business Services: Users, Carriers and Sources of Innovation. In: European Innovation Monitoring System (EIMS), EIMS Publication N° 15
Mittelstrass J (1993) Leonardo-Welt - Aspekte einer Epochenschwelle. In: Kaiser G, Majetovski JF (eds.): Kultur und Technik im 21.Jahrhundert. Frankfurt/New York Campus, pp 16-32
Piore MJ, Sabel Ch (1984) The Second Industrial Divide - Possibilities for Prosperity. New York: Basic Books
Porat MU (1976) The Information Economy. Standford University
Searle JR (1994) Sprechakte - Ein sprachphilosophischer Essay. ed. Frankfurt/M Suhrkamp
Stehr N (1994) Arbeit, Eigentum und Wissen. Frankfurt/M Suhrkamp
Weingart P (1983) Verwissenschaftlichung der Gesellschaft. Politisierung der Wissenschaft. In: Zeitschrift für Soziologie 12, pp 225-241
Werle R (1990) Telekommunikaiton in der Bundesrepublik. New York/Frankfurt/M.
Wiener N (1984) Cybernetics or Control and Communication in the Animal and the Machine. Boston Mass.Institute of Technology, erw.Aufl. 1961

The Way Towards a Knowledge Society - Some Barriers Not Only for Countries in Transition

Lech W. Zacher

1 Toward control of technology – the experience of the West

There are more than C. N. Snow's "two cultures". The first is connected with academia. Scientists produce new visions and ideas, concepts and research results. It is not so difficult to theorize on, say, *information society* or *knowledge society*. There are many outstanding examples (e.g. Bell, Dizard, Castells, Dertouzos). However the approaches applied and conclusions recommended are sometimes divergent. Although it is possible to make some order according to *the level of technocratic thinking* and to *socio-political radicalism*. Technocrats are - not necessarily by definition - technological determinists and technological optimists (Zacher 1981). For them "a bright (information) future is a must". So the future (more often in the singular than the plural) seems to be another "technological fix" to be simply implemented. Technology is perceived *de facto* as an ever growing cake (the more the better?). Some time ago the idea of technological determinism was fundamentally questioned and criticized.

Was this only the idealistic manifestation of humanistically oriented students of science and technology? The choice was promised of at least some social participation in technological decisions. It was a time when the OTA was founded in the USA. So a technology organized "as if people mattered" (to recall Schumacher) seemed to be possible. The ideas of *intermediate technology*, *appropriate technology* and *alternative technology* were elaborated.

So technology seemed almost controllable and effective TA methods were prescribed in the West (see e.g. F. Hetman) and in the East (e.g. G.M. Dobrov - see Chen, Zacher 1978), joint efforts were taken. To sum up: the academic, conceptual and methodological efforts were fruitful and intellectually meaningful.

But besides the academic culture or circle there are also two more. The next circle is connected with decision-making, i.e. with politics. The third culture (or circle or sphere) refers to people, to "parties at interest" (Porter 1991), to citizens, and in some not very distant future to netizens.

Decision-makers of all kinds (political ones, corporate investors and operators, exporters and importers etc.) were not particularly happy about the idea of the *social control of technology* (called e.g. the social assessment of technology, social and environmental impact assessment and the like, see Becker 1997).

Decision-makers felt clearly that technology was important for growth, for international economic competition, for the space and military race during the decades long Cold War. The social control of technology meant less control for government bureaucracies, generals, corporate decision-makers. However technology was too important not to make a political issue out of it. Moreover people in many countries and on an international scale were engaged in antinuclear and environmental protests and movements. There were some spectacular cases of social (or rather political) control of technology, e.g. nuclear test ban treaties, the SST project ban in the United States and the temporary moratorium on genetic engineering. (Needless to add that control of technology - or rather of technologies - extends from research to applications and their impacts, especially long term, unplanned, undesirable, negative side effects). After Harrisburg and Chernobyl people and decision-makers were even more alert. Under democratic conditions there was no way to avoid some change in the relation *power-technology-people*. Political parties in Europe – not only the greens – were introducing environmental issues to their programs. The U Thant Report and then the idea of sustainable development (see especially L. R. Brown's publications) were convincing and promising. Moreover corporations learned that costs of hiding the weak sides of their products and activities and of fighting with consumer groups and activists (like R. Nader for example) and with environmental movements and green-oriented politicians might be very harmful for company image and profits. Often having the best engineers and the best laboratories they were able to modify risky products, to eliminate hard environmental impact technologies and to introduce products and services more friendly to users and the environment. As a result a strong *political will* was finally formed and implemented – some *institutionalization of TA* occurred (in early 70s in the US, much later in Western Europe). The models of institutionalization varied - some more executive branch-oriented, some more legislative, some mixed. Probably the Danish and Dutch experiences were the most elaborated and promising. Anyway there were some institutions, organizations, educational efforts, publications, seminars, research. So some knowledge was cumulated. Needless to state the issue was the accessibility of information (on technologies, their impacts, their interactions) for politicians, for citizens, for consumers. The important role of independent media was evident. In parallel the ideas and concepts of the information society were propounded (by Bell, Masuda).

2 Central and Eastern Europe = hard history, turbulent transition

What happened in the meantime in Eastern and Central Europe? This topic has been extensively treated by many authors, myself included, under labels like "totalitarian state", "fall of Communism", and more recently "transition economies", "democratic transformations" and the like (Zacher 1990, 1994, 1996, 1997, 1998). However for this paper it is just interesting how the three cultures functioned in the area of technology, its decision-making processes, its social and environmental impacts. Of course, there were some principal beliefs, convictions, some ideology connected with technology development. The postwar reconstruction was accompanied by intensive industrialization, urbanization, and militarization. Belief in the Enlightenment *idea of progress* was strengthen by Marxism with its apology for technological progress transformed – by J. D. Bernal – into a theory of *scientific and technological revolution*. However this useful concept was almost totally ideologized and politicized.

Anyway some substantial *modernization level* was achieved (in spite of the Western embargoes on new technologies). Substantial expenditures on R+D and education were secured in state budgets (until the 1980s). However the countries of CEE were (except Czechoslovakia and the GDR) historically less developed than the west of Europe. So in fact they achieved merely what I called a "shallow modernization" level.

Nothing positive can be said on *democracy* and *citizen participation* under totalitarian (or rather authoritarian in the Polish case) regimes. Societies were politically misinformed, censored, media were not free. Any critique was wrong, even in the case of technological decisions. Moreover, there were some dogmas like the one that technology produces bad effects only under capitalism (because of overwhelming profit motivation, lack of planning etc.); also environmental damage was due to capitalist technologies. So the *right to information* was notoriously violated. Belief in central planning and the politicization of everything, technology included, made the debate politically touchy. Specifically *political correctness* emerged even in the R+D sphere where scientists had evidence on of the negative side effects of technology, environmental damages and dangers etc. Information was controlled - censored, restricted, not made public. So this type of society could be called *ex post - information controlled societies.*

The picture was not totally grim however. Some scientists, often engineers, were familiar with Western trends in science and technology assessment. Some preached a "brighter future" which had to emerge from the combination of the results of the scientific and technological revolution and of the "superior socialist mode of production" (see Zacher 1980) while some tried to signal the normative possibilities of technology modification (Zacher 1980). Visions and normative utopias have collapsed. However some positive results from reflection on technology, civilization, future were emerging. *Glasnost* and *Perestroika* helped much tolerable ideas to the expressed and difficult ideas to be articulated. The ineffective economies of "real socialism" were eroding steadily. Then the Solidarity trade union - transformed into mass political movement - started the march towards a more widespread "velvet revolution". Anyway it was now somewhat closer to information or informed society.

Needless to say all revolutions mean destruction (usually they claim - "constructive destruction" to support themselves with Schumpeter's concept). The historical irony was, in the case of CEE, that its societies becoming transitional had "to build" a capitalist (market) system and capitalist society just like in the past they "built socialism". So historical spontaneity was discontinued, the same was true of planned efforts. The introduction of market mechanisms and principles was a priority - economic, political and ideological. The nouveau riche belief in Adam Smith's "invisible hand" has virtually eliminated long term strategies in which science and technology usually play a crucial role. As a result R+D budgets were diminished few times (to less than 0,5% of GDP in Poland). So technology transfer was substituted for technological innovation. This transfer has often come with foreign companies, domestic investment has been low. In fact there was nothing to evaluate. However greater *environmental sensitivity* has emerged. Nevertheless the "shock therapy" (based on World Bank recommendations for troubled countries) emphasized such priorities as short term equilibrium (prices, taxes, finances), privatization (as a source of growth) and deregulation. Restructuring was mostly negative - not supporting state companies (not investing in new industries). Decentralization of the state administration at least in the case of Poland was not only a political reform only but also an instrument for the right wing parties to gain more power. Until now such areas as research, technology, education, culture have been marginalized in state policies. This does not constitute an encouraging framework and atmosphere for technology assessment and citizens' participation (democracy in CEE means only free elections and the freedom to set up political parties).

However it would be wrong to neglect emerging - in spite of everything - elements and signs of the information society. The societies at stake are open to the world, they have free media (to a substantial extent dominated by foreign capital). So the access to information is good. State institutions, banks and companies are more and more computerized. Information technologies are in use, including the Internet (mostly by professionals and young people in schools and universities). For optimists this is a great success, for pessimists - it is still underdevelopment, which is only overcoming the previous "shallow modernization", and is still far from catching up with *high tech societies* which are really information societies. And the passage to a *knowledge society* does not seem not to be a quick move. No doubt growth is a good thing, but how to turn it into development?

A country needs not only entrepreneurial but also educated people (in Poland only around 7% of people have an university education, many are very old, disabled, not productive, or unemployed or have knowledge which is not up to date; the new managers are mostly without higher education). It needs not only capital but strategies as well, not only consumption and commercialization but also social security and culture. There is some hope in the emergence of non-public business schools (already more than 150 in Poland, around 30% of university level education).

Anyway developmental leaps are not likely. However modernization processes may be effectively stimulated provided that the dogma of "invisible hand" omnipotence is replaced by political will, strategies, state and citizens' efforts of all kinds – not only commercial. Of course, the efforts of the private business sector are equally important. Private companies should act as "good citizens". Having this done one can expect more of an "information future", more possibilities for the social assessment of technology (not only justified protests against the construction of highways, buildings, dams etc.), more democracy and citizen participation oriented prospectively. Of course, besides radical and revolutionary changes, many processes are transitory and evolutionary. Anyway there is a great opportunity for innovative policies and entrepreneurial behaviors of the part of governments, business and citizens. In this opportunity is not yet being used of in CEE.

3 Passage to information societies and knowledge based societies - Looking for new ideas

The situation in the highly developed countries is different. Also labels - some more, some less meaningful - are different. These countries are not anymore *industrial democracies*, they are now called *information societies*, *high tech societies*, *digital economies* and the like. Their passage to knowledge-based economies and societies seems to be the "natural" path of development. Apart from being rich and democratic, they are also technological and educational world leaders. They do not need to complete traditional modernization, they are future-oriented. This represents a great *gap* between them and their "poor cousins" from CEE aspiring to NATO, EU etc.

The countries of CEE imitated in their transition the model of the neoliberal economy of the past. This textbook model was never truly implemented since for example in Europe there was for decades quite a strong etatism, welfare state and social market economy, not to mention the public sector and the overwhelming and growing national and EU bureaucracies. Of course, there was a free market but with some planning, strategies, regulations, including by bodies such as the international World Bank, IMF, G-5, OECD, NATO etc. In Japan the role of MITI was crucial for its success. Even in the United States - reading government and corporate documents - it is clear that science and technology (respectively R+D) ought to be priorities if America is to be "number one", so they have to be supported by means of strategies (e.g. in the military and industrial area), special programs (e.g. the Advanced Technology Program led by the National Institute of Standards and Technology, or the Global Information Highway or Y2K). Another priority in the US is education ("preparing the best workforce for the 21st century"). As for technology assessment or rather various forms of *impact assessment,* it can be observed that the concept and methods were adapted to a great extent by government institutions, business, education sector and public (in spite of the fact that Republican-dominated Congress abolished in 1995 the world pioneer - Office of Technology Assessment). There is a growing interest in information technologies and knowledge management. So the path from an information society to a knowledge-based society looks open (but this opportunity is not globalized since there are still the countries in transition and the LDCs).

Neoliberal economics seems not to be adequate to new times and to new challenges. The ideology has also to be modified. New critics and new labels have already appeared. Some, more radical, talk of *technological capitalism* or *information capitalism.* Moreover emerging forms of democracy like teledemocracy, telepresidency etc. are criticized as superficial and manipulatory. Politicians are not happy with the possibility of more *participatory democracy* which can be performed with the help of electronic (information) technology. Is it easier to be a government decision-maker, army general or corporate CEO under conditions of "information noise"?

There are eminent proponents of change in the thinking on growth, the state, globalism, competition and the future (to mention only some: J. Stiglitz, O. E. Williamson, J. K. Galbraith, R. Reich, L. Thurow, P. Drucker, S. Barber, G. Ritzer, M. Castells, the Lisbon Group). Fundamentalist neoliberal economists are *démodé* becoming. Now new ideas and concepts are emerging: postcapitalism, postmaterialism, post-market era, post-information society, and of course knowledge society and knowledge-based economy. Recent debates concern the new role of the state, the new government - business partnership, the new tasks for international organizations, and the new responsibility of leading countries (G-5, G-7) and corporations, especially the transnational and global. Something is happening in the West. It proves that the commercialization of minds has not been total. Thank God. Perhaps the world will not enter a new Renaissance or Enlightenment, but it may avoid neobarbarism. However the latter requires some effort to make societies not only information but *informed* ones, and to treat knowledge not only as a production factor or instrument of power, manipulation and domination but as a value *per se*.

4 Some concluding remarks

To sum up: the possibilities, opportunities and challenges of the information age vary from country to country, from region to region. Less developed countries or Central and Eastern Europe do not have the same chances as the world leaders in science, technology and education. Moreover they are the subjects of economic expansion and of the brain drain. Their ideology of development is narrow and shortsighted, oriented mainly to equilibrium, not to change and the future. Emerging information and knowledge-based economies and societies may widen the gap. If so, this would be unfortunate - *parallel worlds* may emerge. This may cause situations of conflict on a domestic, regional and global scale. For the sake of future global security existing disparities and exclusions should be fought against. And this is not a case for a technological fix.

References

Becker H (1997) Social impact assessment: method and experience in Europe, North America and the Developing World. London, UCL Press Ltd.

Castells M (1996, 1997, 1998) The Information Age: Economy, Society and Culture. vol. 1 The Rise of the Network Society, Vol. 2 The Power of Identity, Vol. 3 End of Millennium, Blackwell, Oxford

Chen K, Zacher L (1978) Toward Effective Technology Assessment. In: Dobrov GM, Randolph RH, Rauch WD (eds.) Systems Assessment of New Technology: International Perspectives. IIASA, Laxenburg

Kukliński A (ed.) (1996) Production of Knowledge and the Dignity of Science. Warsaw, Euroreg

Kwiatkowski S, Edvinsson L (eds.) (1999) Knowledge cafe for intellectual entrepreneurship. Warsaw, Leon Kozminski Academy of Entrepreneurship and Management

Martin C (1999) Net Future. New York, Mc Graw Hill

Porter A et al. (1991) Forecasting and Management of Technology. New York, John Wiley

Zacher L (1980) Premises and Goals of TA in Centrally Planned Economies. In: Chen K, Boroush M, Christakis AN (eds.) Technology Assessment: Creative Futures. Elsevier North Holland, New York

Zacher L (1981) Toward Democratization of Technological Choices. Bulletin of Science, Technology and Society, Pergamon Press, vol. 1, no. 1-2

Zacher LW (1990) Institutionalisierung von TA in einem postkommunistischen Land - der Fall Polen. Kassel Universität, Oktober

Zacher LW (1994) Social Movements as a Part of the TA Process (Unrealized Possibility). In: Zacher LW (ed.) Understanding the Contemporary World - Inquiries into the Global, Technological and Ecological Issues. vol. 1, Warsaw 1994, Educational Foundation TRANSFORMATIONS

Zacher LW (1995) The Unrealized Vision of Future: The Case of Radowan Richta Predictions. Dialogue and Universalism, vol. 5, no. 11-12

Zacher LW (1996) Chances of Modernization in Post-Communist Countries. Journal of International Studies, vol. 4, No. 3, Summer

Zacher LW (1997) Poland: technology problems in a typical transition economy. In: Dyker DA (ed.) The Technology Policies for Transition Countries. Budapest, CEU

Zacher LW (1998) Modernization in Eastern Europe, and Post-modern Restructuring in the West: Looking for Compatibility. In: Manniche J (ed.) Searching and Researching the Baltic Sea Region. Bornholm, Research Centre of Bornholm

Problems of the Production and Application of Knowledge in a Societal Transformation. The Czech Republic's Approach to Knowledge Society

Stanislav Provazník, Adolf Filáček, Petr Machleidt

1 Introduction

The stage into which the world economy has recently entered is marked by a dual and profound transformation. First: under the impact of a considerably liberalized international market, national and local economies are fast changing into a global economy. Secondly, the importance of each country's traditional production factors has been sharply declining; on the contrary, an ever greater role is being played by each country's economic capacity to produce technological innovations, an ability which is based, on the one hand, on the upswing of free enterprise, and on the production and utilization of knowledge, on the other. Knowledge is thus becoming a vital instrument in concentrating practical activities when tackling major social issues and environmental challenges.

In their interaction, both key contemporary processes mentioned above tend to be multiplied, with the classical industrial criteria receding into the background under their joint impact; on the contrary, the developmental dynamic, based on people's competence, knowledge and creativity, on the progress of scientific knowledge, on the abilities of the entrepreneurial actors to re-organize continually both human and knowledge resources, on technological efficiency and innovation abilities, seems to be asserting itself virtually anywhere. In their entirety and interconnection, these megatrends characteristic of the current societal development patterns are perceived as paths leading to a society which establishes its reality and its future on the sources of knowledge, thus differing from the industrial society. An analytical contrast between the industrial society and the knowledge society as theoretical models representing two basic processes of the modern world is helpful in capturing – in conceptual terms – the stream of turbulent changes to which we are daily exposed. The intricate empiricism of our days is becoming clearer, more transparent, and – consequently – bearable, when shown to be an outcome of projections, transitions and mediations of two divergent and outright contradictory processes. However, the key point is that, while using such models, we are in a position better to observe what has already emerged, what is being established, that we are no longer letting our future slip away through our fingers, so to say. For each individual country, these processes, conducive to the formation of a society in which the production and application of knowledge will be, or has already been, determining people's lives and work, assume an objective character: the only question concerns the actual mode, nature and rate of intensity of its involvement.

However, understanding this concept in the Czech Republic is far from common, and the term "knowledge society" is still frequently questioned. This can be partly explained by the Czech society's aversion to such "pure concepts", to "distinct patterns", such as "industrial civilization", "industrial production system", "science and technology revolution", "socialism", "capitalism" etc. Indeed, this country has a history in operating with such typological terms in its political practice in the past few decades, a practice that often served to subordinate empirical reality to its own logical models, and, by employing a priori subjective ideological intention, politicians were free to designate certain historical facts as significant, dismissing others as insignificant, while abstracting from those empirical phenomena which appeared to interfere with the ideal (in fact ideologized or ideological) image of reality or the logic of a historical process. While using the term "knowledge society" today, we should certainly respect this negative historical experience, paying special attention not to place the logical before the historical, to make sure that empirical reality is no longer deduced from concepts and wishful thinking, to ensure that statements pertaining to concepts and models are no longer stylized as statements on realities; in other words, the knowledge society should not appear solely as an abstract and distant vision, it should never prevent us from grasping those dimensions of reality it has previously failed to incorporate into its own vision and its real empirically recognizable content should be precisely defined.

Once we have succeeded in clarifying this particular methodological issue, and set our sights on supporting the general models through a specific historical analysis, we are likely to encounter yet another obstacle in the Czech society; namely the widespread view that the issue of the knowledge society is not topical for the country's transforming society, and that the very efforts to delineate this problem tend to divert the society's attention from what are seen as more important and immediate challenges. To a certain extent, this opinion may be justified by the country's contradictory reality. The ongoing economic reform, which has been successful in many respects, is – at the same time – accompanied by an unexpected, exceptionally steep and persisting decline in the country's GNP. This situation makes it imperative for the society to deal with immediate and urgent issues, face an economic crisis and mounting social tensions, and stipulate – in no uncertain terms – how much energy and resources should be devoted to tackling long-term objectives, how much strength should be spent on solving long-range problems today and which particular problems should be left to be tackled in the future. This problem relating to long-term and immediate tasks and goals, which spells out the genuine contradiction arising out of the specific course of social transformation, would certainly call for a more thorough analysis. However: features indicative of the knowledge society are already present in our reality.

2 The Structural Problems of the Transformation Concept

True to say, thanks to the institutional changes (privatization, price and foreign trade deregulation etc.), the Czech Republic's transformation process so far has paved the way for a decisive economic re-orientation. At the same time, the past few years have seen an accumulation of the country's economic problems which, in their entirety, suggest that the Czech society has come up against a still deeper layer of the economic process, a stratum that has not been previously reflected in the existing overviews of the transformation process: all the signs are that economic processes taking place even within the society's new institutional structures of the market economy have been reproducing mostly those proportions based on the perpetuation of the classical criteria of the industrial society and on an absolutely inadequate application of a modern developmental dynamic, proceeding from a structure connecting business activities with scientific and technological knowledge into a well-functioning national innovation system.

While in the current advanced economies the classical criteria of the industrial society have been receding into the background, and knowledge has been gradually becoming the most decisive factor of the whole economic process, while ever more intense world-wide economic competition has been bringing pressure to bear on the advanced countries to offer more and more products and services that contain a large amount of knowledge, the competitive advantage of the Czech economy within international economic competition still displays a largely different nature and different parameters. In its entirety, the Czech economy is today based primarily on low wages of its work force and on less qualified labour. Such a strategy leads neither to the improvement of the position of the Czech economics in the international competition, nor to the accelerating of the GNP-growth and productivity of labour. Its result is, on the contrary, a deterioration of the national industrial structure.

Even though the overall wage level (measured by the average gross wage in the official exchange rate) has, in the past few years, been depressed to the lowest point among Central European countries (with the exception of Slovakia whose wage level is yet lower), this measure has still failed to result in any improvement in the Czech Republic's position in international economic competition. To illustrate this point, it is sufficient to make the following comparison:

- as compared with the average figures in the EU countries, the Czech Republic amounts to slightly over 50% of their labour productivity as well as their wages (in terms of the purchasing power parity); and this gap has not been narrowed during the years of economic transformation in the Czech Republic; what is worse, the gap has rather been growing wider in the past few years;
- as compared with its post-communist neighbours - Poland and Hungary - the Czech economy is currently accounting for just 50% of their growth-rate of their per capita GNP as well as for 50% of the growth-rate of their labour productivity.

The same picture emerges from the basic data describing the structure of the economy. While in recent years, the industrially advanced OECD countries have made the production of goods and services with a high level of added value the backbone of their economic performance and international competitiveness, and more than half of their GNP is now generated in production plants based on high technology and a high level of knowledge, the Czech industry has, on the contrary, been registering a slump in the production of industrial branches noted for a higher rate of invested qualified labour. Indeed, the share of the production of machinery and equipment, electric and optical instruments and of the engineering and electro-engineering industries in general in the national total industrial output has been declining, and the overall ratio in the number of companies with high-, medium- and low- level of technology is estimated at approximately 15% : 60% : 25% respectively. Over the past ten years, the export volume of those manufacturing branches in the advanced OECD countries has more than doubled; in the Czech economy the share of high-tech products in the overall exports in 1996 totalled a mere 6.3%.

A profound analysis, therefore, clearly shows that the economic development pattern based on low wages and low qualification of employees, i.e. typically industrial trends, has exhausted its potential in the conditions of the Czech economy, and no longer leads to any increase in the country's competitive edge. The Czech Republic is a country with the long-standing industrial traditions, considerably high education level of its work force and relatively high living standards and can hardly win the race for the cheapest labour. If the country continues to base its position in international economic competition on low wages, it is likely to start moving ever faster on a downward spiral of living standards as foreign companies which give preference, in their investments, to lower wages are certain to find – even in the Czech Republic's vicinity – countries offering yet cheaper labour for a long time to come. Therefore, the Czech Republic faces in the near future a crucial task, namely very fast to aim at a major change in its economic orientation and to base its competitive advantage on the growth of its labour productivity and the ability to produce innovations, on the production and application of knowledge, i.e. trends characteristic of the knowledge society. Without mastering it, the societal transformation cannot be successful.

3 The Role of Human Resources (Human Capital) in Economic Transformation

The available extensive body of literature studying the phenomena relating to the emerging knowledge society in the 1990s shows that the new reality introduced by the advance of the trends typical of the knowledge society is confined neither to a one-dimensional impact of modern technology (using massive capital investments), nor to a high growth-rate of scientific and technological information. Still, running beneath the surface of these changes, which – in their own right – constitute external features of the emerging knowledge society, there are the following interrelated changes in all three key components of the development of the society's scientific and technological potential: i.e. (a) in the system of knowledge production, especially in the national research potential; (b) in knowledge application in the level of technological facilities; (c) in knowledge application in connection with the process of activating human resources.

Contemporary developments seem to assume the necessary dynamic solely if and when all these three components are harmoniously developed. An ever more prominent role is being played in this process by the subjective factor, by man in the full range of his skills and talents, a component also called human capital. The link between economic efficiency and human resources, or rather human capital, is growing to be a key proportion of today's economy. That is why, ever greater importance is ascribed to the introducing such methods of economic stimulation and human motivation which, in addition to instigating higher rates of productivity, also make it possible to cultivate human resources, develop that source known as the human capital. Seen in this light, the rise of the knowledge society and its requirements represent a major turning point in society's production parameters.

In the past, human potential in education, science and technology figured prominently as the best developed part of the potential of the Czech society. Thanks to its free education at all levels, this country has had a relatively high share of well-educated population. But the previous totalitarian regime failed to make an adequate and sufficient use of this particular potential, which has not been fully actualized until this day. On the other hand, this human capital should be assessed quite soberly, while shedding any unsubstantiated illusions about its strength. Such illusions have accompanied societal transformations in all East European countries, stemming from the fact that each country's potential of the available human capital is evaluated according to static and formal yardsticks, regardless of the actual prerequisites of its application in the demanding conditions of international market competition.

The most efficient source of the human capital are well-educated professionals who are capable to produce and apply new research results. Set in a spiral of the increasing supply of new knowledge – which keeps stimulating their application, giving rise to yet new supply at a still higher level, the process of producing knowledge assumes its own dynamic, which is typical of the present-day era and which accelerates social changes conducive to the knowledge society. The integral part of the transformation process in the Czech Republic is its sweeping educational reform, which has already found its positive expression in the quantitative developments of the educational system, and in changes of its structure. In the 1990s, the number of Czech university students has grown by more than 40%, the number of enrolled applicants soaring by more than two thirds. The number of faculties has increased by over a third. The Czech Republic seems to be moving from an elitist university system to a mass model. Yet, as compared with OECD countries, the number of the university students at Czech higher education institutions and the share of the university graduates in the overall population is still quite low. On average, Czech universities still take only about one half of all the applicants. A remarkable aspect is the upsurge of the country's secondary vocational and secondary comprehensive educational systems. In the 1990s, the number of secondary vocational schools, which boast of a long tradition and considerable prestige in the Czech Republic, has more than doubled, the number of their students growing almost by half.

On the other hand, today's Czech approaches to education and to the cultivation of human resources often hark back to that stage of industrialization during which no special accent had to be placed on the promotion of the existing human resources. Up till now, the Czech Republic's political and economic scene has failed sufficiently to reflect the radical changes that have occurred in the world economy during the 1990s, namely the fact that the management of human resources and their preferential development has become an absolute priority.

Seen in this context, the most important problem is that neither the structure of production, which is strongly influenced by the Czech Republic's current position in international trade, nor the structure of its economic interests, arising out of the fast but inconsistently completed privatization process, has yielded enough incentives to raise work performance, the educational and qualification levels of the country's labour.[1] As a result, a considerable portion of the work force is not directly bound, in their incentives, to the growth in their productivity and innovation progress. This situation is instrumental, among other things, in maintaining too much useless and unskilled labour in Czech industrial plants. Furthermore, the fact that the country's generous welfare system is sometimes detached from performance-related parameters is hampering the task of forging positive links between work performance and one's social position.

Another problem is the slow pace of the Czech Republic's industrial modernization and restructuring, caused by a shortage of investments. This results in ageing industrial production technology and equipment which, in turn, limits the possibilities of human resources management and increases the delay in applying and utilizing knowledge throughout the society.

A similar problem is caused by a shortage of investments into education and human services. The current system of professional development is based on the resources and human capital reserves created in the past. Due to the current low dynamic of the development in education, medical services and research, these branches which are supposed to pave the way for future economic progress are rather lagging behind, being financially undernourished.[2]

4   The Innovation Function of Research

Let us now take a closer look at the third key component of the society's scientific and technological potential – the system of research. Throughout the transformation process, the research system has been restructured in keeping with the principles still valid in the other European democracies. A salient feature of that process was an effort to preserve the system's valuable scientific and research

---

[1] Another move in the same direction is the influx of workers into the elementary service sector whose level and culture of work is often low. Due to its large capacity, this sector has actually prevented the labour market from being fully developed so far.

[2] Although in terms of the share of its education spending in its GNP the Czech Republic is roughly on a par with the average OECD countries, in terms of its per capita spending (broken down per one secondary school student and one university student), our country finds itself well below this average, earmarking approximately only half of the expenditures of other OECD countries.

potential. This key and generally accepted goal of the transformation process has been implemented to a varying extent in different research sectors.

Industrial research has met the most dramatic fate of all the research branches. Generally speaking, its transformation was guided by market forces with marginal support from the government. That fact together with an unfinished ownership transformation and the faltering process of restructuring many production plants has been instrumental in the following situation: the Czech Republic's industrial research has seen not only the biggest personnel reductions but also suffered relatively considerable losses in its research potential. Some research areas have found themselves on the verge of liquidation. What is still worse, former contacts between enterprise-based research and fundamental research in the Academy of Sciences and the universities have been loosened.

The transformation process of research at the higher education institutions is long overdue. It is especially the process of diversifying and identifying valuable and less valuable research (valuable research being positively assessed research by international evaluation boards) that has yet to be finished. Even though the cuts in the number of people engaged in research have been quite modest, due to various causes there has been no appreciable rise in scientific research potential so far.

The Academy of Sciences of the Czech Republic has been the only sector which, despite far-reaching personnel reductions, has not only come out of the transformation process preserved but also partly strengthened. The Academy of Sciences has not only willingly opened itself to international evaluation and competition. It has succeeded in exploiting the current restrictions of its financial resources in order to raise the productivity of its scientific research, which is a quite unique accomplishment.

Seen in terms of an international comparison, the entire transformation process in the Czech science and research sector appears in favourable light. Following the radical cuts in the state research funding in the 1990s, the Czech scientific and research community did not resort to a defensive, "survival strategy" in a bid to save the existing research institutions and researchers. Instead it actively anticipated society's requirements towards science and research, and accepted the internationally acknowledged norms of evaluating research quality. Following on from there, it has meanwhile refocused and restructured its research projects, reducing what were seen as inferior components of its research potential as well as those which proved to be difficult to adjust under the new circumstances.

Needless to stress perhaps, during the financial difficulties accompanying the transformation the Czech government considerably reduced its science and research expenditure, which has become the main, and – in the case of academic and university research – virtually the only available source of funding. But the fact is any neglect of investments into the future developments should never assume dimensions actually threatening the country's future. Nevertheless, there are several processes in the Czech Republic that seem to be pointing exactly in that undesirable direction. Meagre government subsidies channelled into science and research, persistent shortages of funds to be used by industrial plants for their own and contractual research, and last but not least the decreasing possibilities to obtain foreign grants – all this has lead to unjustifiable restrictions of investments into the country's research infrastructure, to the actual impossibility of paying decent salaries to scientific and research personnel and to a sharp decline in the attractiveness of the scientific career for young talented people. Here are two examples to illustrate this point:

As compared with 1989, the real value of the country's overall research and development (R&D) spending has now declined to 28.2% of the total, falling to 73.6%, when broken down per one university-educated employee in research and development.

The real value of average monthly income of people employed in the research and development sector reached the 1990 level as late as in 1996 (today its index accounts for roughly 110% of the 1990 figure). The growth-rate of the average monthly income in the Czech science and research establishment has just kept pace with the growth-rate of the average monthly income in the Czech Republic. The former was, however, kept well below the growth-rate of the average monthly income in many other branches. For instance, as compared with the actual growth-rate of the average monthly salary in public administration, since 1992 the growth-rate of the average income in Czech R&D has been annually between 10% and 15% lower than in state administration, its real value currently accounting for only 93% of the average income of a civil servant. Back in 1990, the average monthly income in the Czech research and development sector was, on the contrary, 10.3% higher than that in public administration.

We may conclude by saying that while the completed legislative and institutional changes in the country have managed to promote the autonomy and independence of Czech science, opening a broad scope for action to scientific and research institutions as well as researchers, the current economic situation and the related financial constraints have again visibly narrowed down that space. Still, research institutions are in a position to respond to the present-day problems: they can reduce their research and engage instead in commercial activities to earn desperately needed financial resources or they can make further cuts in their creative potential, which has been already selectively reduced anyway. The chances are that either course could limit the scope of the Czech research sector to such an extent as to substantially jeopardize its overall quality.

5   Key Position of Innovation Policy

The overriding trait of the current innovation system in the Czech Republic is that – despite the difficulties stemming from its contemporary economic problems - it still has a scientific, research and technological base whose key features are similar to those in the advanced Western countries. These include a combination of objective (technological and industrial facilities) and subjective (cultural, intellectual, emotional) prerequisites of the process of generating and applying knowledge, including a well-elaborated system of research institutions. The primary task is to preserve this base during the period of transformation difficulties and to develop it.

As a result of the prevailing contemporary megatrends, there have also been changes in the competitive advantages of individual countries. Global competition has been laying greater claims on the competitive skills of not only companies but also whole states as the innovation potential of today's companies does not depend solely on the abilities and skills of individual businessmen but, to a growing extent, on each country's knowledge-based, research and general institutional infrastructure without which individual entrepreneurs could hardly hope to do successful business at all. That is also a new phenomenon of the past decade.

At present, each country's competitive advantage is not given once and for all. It is rather dependent on efficient national innovation policy (including, among other aspects, a science and research policy, which thus assumes extraordinary importance). Such a national policy is geared to manage society's innovation sources and build a national innovation system in a well-thought out manner. In order to promote such a policy it is necessary to develop truly productive and innovative enterprise sector as well as finish the construction of a competent and well-functioning national research and development base, and upgrade the quality of the nation's education system.[3]

The new concept of the Czech Republic's innovation policy makes it imperative to adopt a new approach to mutual contacts between universities, research institutions and industrial plants, a type of co-operation which is nowadays absolutely unsuitable and which is not supported by any legislation either. It introduces anew the issue concerning the diffusion and commercial distribution of research results and the task of transcending institutional boundaries between the production of knowledge and the knowledge market. And finally: the issue concerning the generation of knowledge: it should be reformulated in global competition and international networks as well.

Quite a disturbing aspect of the current situation in the Czech science and research sector is the prevailing official view that the actual scope of its research should be adjusted to match the current unsatisfactory state of the Czech industrial production, that we should reconcile ourselves to its present-day slump since the overall funding of research and development reflects the country's economic possibilities and is consonant with the relative amount of science and research spending in the countries of the European Union.

Research and development in the Czech Republic is currently funded by roughly 1% of the GNP. It is relatively comparable with the expenditures of smaller, less developed EU countries, such as Greece and Portugal which do have a similar per capita GNP but whose historical, social and economic traditions differ from the traditions of the Czech lands. On the contrary, countries with comparable traditions and production structures are known to earmark to research and development a much higher share of their GNP: Austria - 1.52%, Netherlands - 1.89%, Denmark - 1.8%, Finland - 2.22%, Norway - 1.94%, Sweden - 3.26%,

---

[3] Nowadays, the importance of the first factor is beginning to be understood but the significance of the second one still remains underestimated. J. Schumpeter who so vehemently justified the key role of the businessman for that "creative destruction", which forms the basis of the dynamic of a modern market economy, emphasized that a businessman operates in a space delineated by two different poles: on the one hand by laws and the ethos of the capitalist process, which enables the businessman to retain, at least for a time, the advantages of the new technology he has created; on the other hand this space is outlined by the resources of scientific knowledge, a pool from which the businessman's research draws. During the revival of Schumpeter's ideas in the Czech Republic, that second source of free enterprise, which is nurtured from the social infrastructure, seems to have been forgotten. Naturally, research results are not needed for the speculative, non-productive type of business. (s. Schumpeter 1950, pp180-181).

Switzerland - 2.68% respectively. In terms of their science and research spending, an absolute majority of small European countries are either around or above EU average.[4] This position is correlated with the share of high-tech products in the overall industrial output of those countries. That too amounts to roughly the European Union's average. A further detailed analysis of these figures seems to suggest that the average R&D spending in EU countries, measured as a percentage of their GNP (i.e. 2%), constitutes the bottom line verified by practical experience which is vital for high-tech development of the advanced industrialized European countries.

The key conclusion ensuing from this paper is as follows: support for research and development is not solely the function of the level of economic activities at any given time. It depends on many other factors, especially on national economic structure, its share in world trade, traditions of science and education in its society etc. Last but not least, it depends on the decision of the society itself whether it really wants to set out on the road of innovation with all the positive economic and social consequences stemming from such a decision or whether it is prepared lower its own intellectual, educational and knowledge-based standards and rather concentrate on mass production and cheaper labour with a lower amount of intellectual creativity.

6   Appendix

The following tables and figures illustrate how the rise of the "knowledge society" appears in the Czech Republic, a country that has been undergoing a wide-ranging transformation since the early 1990s. During that process institutional, cultural and value-related prerequisites have been gradually established for the purpose of introducing and operating a democratic state with a market economy, a system well-integrated into Europe's political and economic structures. The tables and figures also demonstrate to what extent the emerging knowledge society has been modified by the ongoing social transformation and how, on the other hand, does this nascent knowledge society itself affects the problems and tasks facing social transformation.

---

[4] A yet more convincing argument in this context is a comparison of per capita R&D spending in the OECD countries. In this comparison, all the afore-mentioned European countries are above the OECD average, and - with the exception of Austria - above the EU average as well. There is only Turkey trailing behind the Czech Republic, provided that 1 US dollar is calculated according to the exchange rate: 1 USD totals 30 CK. If, using the current purchasing power parity rate 1 USD = 12 CK, also Portugal and Greece would finish behind the Czech Republic. However, the exchange rate according to the purchasing power parity is known to distort the real situation, particularly so because the Czech science and research sector has been buying its instrumentation and information technology for prices fixed within the framework of the official exchange rate.

Table 1: Czech Republic – research and development aggregate indicators; Source: Statistical yearbook of the CZ, 1993-1999

	1989	1990	1991	1992	1993	1994	1995	1996	1997	1998
Employees in R&D sectors - total (31 December)	137.927	105.916	76.487	57.227	40.214	38.752	47.500	49.632	52.245	51.198
Including: R&D employees	77.850	62.268	41.668	31.543	23.336	23.741	37.151	41.544	44.744	45.557
- Including:										
- with university education	46.279	31.248	23.277	18.821	13.266	13.749	15.941	17.706	18.520	19.003
- with full secondary and higher education	28.832	20.782	*)	11.768	9.278	8.923	13.874	14.232	14.734	15.888
- R&D employeees - scientists	9.388	7.855	6.381	5.923	5.015	5.610	9.255	9.606	11.490	10.666
Total non-capital resources for science and technology (CZK million)	19.583	10.763	13.503	12.204	11.091	11.215	12.431	14.031	16.870	20.136
Annual change in %		-45,0%	25,5%	-9,6%	-9,1%	1,1%	10,8%	12,9%	20,2%	19,4%
Including:										
- From state budget		3.078	3.451	2.654	2.542	2.982	3.951	5.077	6.359	7.219
Annual change in %			12,1%	-23,1%	-4,2%	17,3%	32,5%	28,5%	25,3%	13,5%
Total capital expenditure on science and technology (CZK million)	1.837	1.652	1.708	2.295	1.229	1.768	1.551	2.226	2.607	2.729
Annual change in %		-10,1%	3,4%	34,4%	-46,4%	43,9%	-12,3%	43,5%	17,1%	4,7%
Total expenditure on science, research and technology (CZK million)	21.420	12.415	15.211	14.499	12.320	12.983	13.982	16.257	19.477	22.865
Annual change in %		-42,0%	22,5%	-4,7%	-15,0%	5,4%	7,7%	16,3%	19,8%	17,4%
Gross domestic expenditure on R&D in % of GDP	4,08%	2,14%	2,03%	1,71%	1,21%	1,10%	1,01%	1,03%	1,16%	1,26%
Expenditure on one R&D employee (CZK thousands)	155,3	117,2	198,9	253,4	306,4	335,0	294,4	327,6	372,8	446,6
Expenditure on R&D per 1 inhabitant of Czech Rep. (CZK)	2.067	1.198	1.476	1.405	1.193	1.256	1.353	1.576	1.890	2.221
Expenditure on R&D in $ per 1 inhabitant (var. 1 $ = 30 CZK)	69	40	49	47	40	42	45	53	63	74
Expenditure on R&D in PPP $ per 1 inhabitant (var. 1 $ = 12 CZK)	172	100	123	117	99	105	113	131	158	185

Table 2: Domestic expenditure on research and development in the Czech Republic: by sector and source of financial means [in CZK thousand]; Source: Statistical Yearbook of CZ, 1999

Sector	1995	Structure in %	1996	Structure in %	1997	Structure in %	1998	Structure in %
Total domestic expenditures on R&D - source of financial means								
1. Business enterprise sector	8.824.092	63,1%	9.693.413	59,6%	11.651.646	59,8%	13.762.848	60,2%
2. Government sector	4.513.020	32,3%	5.776.861	35,5%	5.996.688	30,8%	8.423.316	36,8%
3. Higher education sector	137.325	1,0%	445.029	2,7%	1.410.358	7,2%	69.088	0,3%
4. Private non-profit sector	40.606	0,3%	28.187	0,2%	47.863	0,2%	13.815	0,1%
5. Rest of the world sector	467.567	3,3%	313.376	1,9%	370.887	1,9%	595.903	2,6%
Total	**13.982.610**	**100,0%**	**16.256.866**	**100,0%**	**19.477.442**	**100,0%**	**22.864.970**	**100,0%**
In % of total R&D expenditures	*65,1%*		*59,9%*		*62,8%*		*64,6%*	
Business enterprise sector - source of financial means								
1. Business enterprise sector	8.364.534	92,2%	8.832.481	90,7%	11.095.698	90,7%	13.197.021	89,4%
2. Government sector	406.707	4,5%	714.645	7,3%	962.279	7,9%	1.207.780	8,2%
3. Higher education sector	2.900	0,0%	2.868	0,0%	3.776	0,0%	0	0,0%
4. Private non-profit sector	17.917	0,2%	22.614	0,2%	3.140	0,0%	9.698	0,1%
5. Rest of the world sector	286.567	3,1%	166.993	1,7%	167.211	1,4%	345.306	2,3%
Total	**9.098.625**	**100,0%**	**9.739.601**	**100,0%**	**12.232.104**	**100,0%**	**14.759.805**	**100,0%**
In % of total R&D expenditures	*65,1%*		*59,9%*		*62,8%*		*64,6%*	
Government sector - source of financial means								
1. Business enterprise sector	416.424	11,3%	851.057	16,8%	504.679	9,7%	481.089	8,2%
2. Government sector	3.128.559	84,6%	4.107.655	81,1%	4.503.918	86,8%	5.265.383	89,6%
3. Higher education sector	21.379	0,6%	17.807	0,4%	46.500	0,9%	5.233	0,1%
4. Private non-profit sector	19.396	0,5%	3.795	0,1%	41.458	0,8%	2.452	0,0%
5. Rest of the world sector	112.376	3,0%	82.595	1,6%	91.061	1,8%	123.657	2,1%
Total	**3.698.134**	**100,0%**	**5.062.909**	**100,0%**	**5.187.616**	**100,0%**	**5.877.814**	**100,0%**
In % of total R&D expenditures	*26,4%*		*31,1%*		*26,6%*		*25,7%*	
Higher education sector - source of financial means								
1. Business enterprise sector	23.134	2,0%	6.176	0,4%	26.540	1,5%	44.367	2,0%
2. Government sector	977.754	82,5%	945.640	65,7%	526.517	29,6%	1.933.443	89,1%
3. Higher education sector	113.046	9,5%	424.034	29,4%	1.110.484	62,5%	63.855	2,9%
4. Private non-profit sector	3.293	0,3%	922	0,1%	1.889	0,1%	1.665	0,1%
5. Rest of the world sector	68.624	5,8%	63.544	4,4%	112.615	6,3%	126.940	5,8%
Total	**1.185.851**	**100,0%**	**1.440.316**	**100,0%**	**1.778.045**	**100,0%**	**2.170.270**	**100,0%**
In % of total R&D expenditures	*8,5%*		*8,9%*		*9,1%*		*9,5%*	
Private non-profit sector - source of financial means								
1. Business enterprise sector	-		3.699	26,3%	24.729	8,8%	40.371	14,4%
2. Government sector	-		8.921	63,5%	3.974	1,4%	16.710	6,0%
3. Higher education sector	-		320	2,3%	249.598	89,2%		
4. Private non-profit sector	-		856	6,1%	1.376	0,5%		
5. Rest of the world sector	-		244	1,7%				
Total	-		**14.040**	**100,0%**	**279.677**	**100,0%**	**57.081**	**100,0%**
In % of total R&D expenditures			*0,1%*		*1,4%*		*0,2%*	

Figure 1: Czech Republic - Gross Domestic Expenditure on R&D in % of GDP; Source: Statistical Yearbook of CZ, 1999

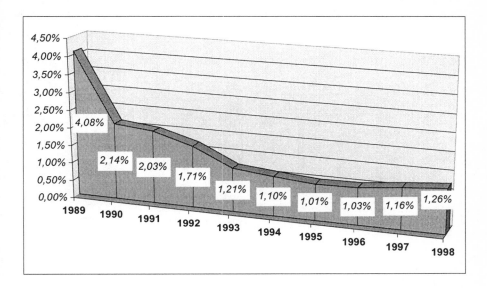

Figure 2: Czech Republic - labour force structure by sphere (national economy total = 100 %); Primary sphere: agriculture, hunting, forestry, fishing, mining and quarrying; Secondary sphere: industry (excl. Mining and quarring), construction; Terciary sphere: all others branches of CZ-NACE; Source: Statistical Yearbook of CZ, 1999

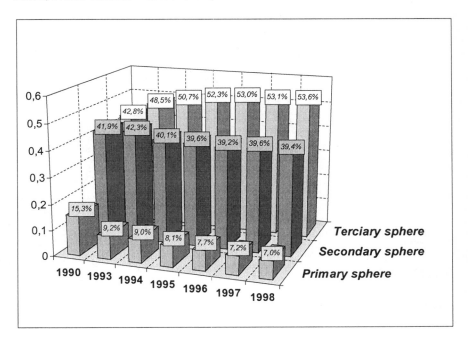

Figure 3: Czech Republic – Real Value of Average Monthly Salary in R&D (1989 = 100);
Source: Calculated from data of the Statistical Office of CZ, 2000

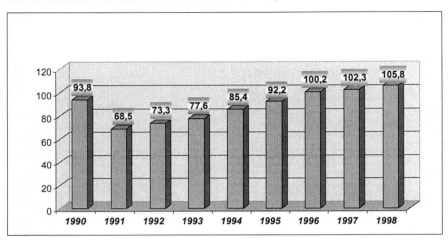

Figure 4: Czech Republic – employment in selected branches (civil sector of national economy =
100%); Source: Statistical Yearbook of CZ, 1999

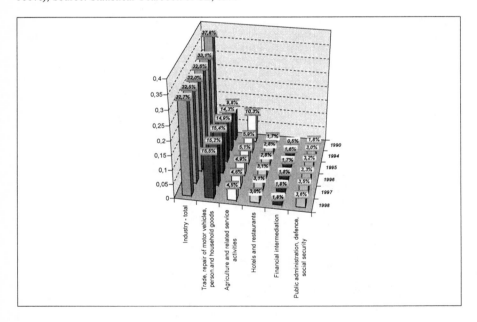

References

Schumpeter J (1950) Kapitalismus, Sozialismus und Demokratie. A. Francke AG Verlag Bern

Innovation and Financial Intensity of R&D in Poland as the Polish Approach to the Knowledge Society

Ewa Okoń-Horodyńska

> "The new dividing line is not between the haves and have-nots. It's between the 'knows' and 'don't knows'"
> (Klaus Schwab, the founder of the World Economic Forum in Davos)

1   Introduction

So-called "adjustment process", one leading to integration of Poland with European Union, and fulfilling the successive conditions arising from pre-accession agreements are issues which have been the focus of Polish scientists and politicians attention for a few years. However, in the masses of questions concerning defence of interests of Polish industry, trade or agriculture, we forget about the basic problem, i.e. adjusting level of knowledge, institutions developing and storing it up, which finally provide conditions for the production characterised by high concentration of knowledge - a guarantee of high quality. Relationships between Europe and Poland in the fields making possible yielding and accumulating can be cursorily presented as follows. In Polish science we have had (both historically and currently) many achievements of European and world importance. Universities and higher schools in Poland, however not financed sufficiently, educate students on high level, students who easily find jobs all over the world. It is different with modern technologies - but it is assumed that market economy development in democratic society open to international competition will lead to transfer of innovations in technology based on concentrated knowledge. Dissemination of products of high content of knowledge is also a way to knowledge society.

The subject for consideration in the article is an analysis of R&D area which should be responsible for making products of concentrated knowledge. The analysis was made for Polish economy which is undergoing system transformation in all its sections. A thesis is submitted here that quality, role and institutional assistance for R&D area in an economy are substantial contribution to the process of creating complex knowledge society.

2   Activity of R&D area in Poland
(now undergoing transformation) - state diagnosis

The R&D section in each economy belongs to the area of widely understood knowledge enabling production characterised by its high concentration. The effects can be generally accessible or the access can be for different reasons limited. Part of R&D section output are publications which accumulate knowledge and disseminate its range in society. However the most important part of R&D is research and its findings because it is potentially connected with innovations. In the period before transformation a specific system of scientific research established in Poland. On the one hand the state supported science which was regarded as the foundation for economic and social development. On the other hand no institutions connecting science and economic were developed. As the main part of scientific activity was regarded the activity grouped in separate section called "Science and Technology Development". Because Science was treated as a separate section it resulted among others things in determining the content of statistical data concerning science as they almost completely concerned the internal organisational characteristics of the science section. All the activity of R&D performed outside the "Science and Technology" section was treated as being outside the system. Similarly costs and results of R&D for the benefit of the army were not included in R&D statistics. The way R&D section was functioning in different economic systems is also impossible to compare. Only those several differences made it impossible to compare costs and effects of R&D section in countries of different economic systems. Thus attempts towards comparative analysis require not only purely statistical approach but also extensive descriptive assistance. Relationships between science and industry in former socialistic countries resolved into obligatory forcing of findings into companies which because of lack of competition were interested in technological progress only to a small extent. There were also three rigid divisions of science, i.e. Polish Academy of Sciences holding a monopoly on basic research, branch institutes and research-and-development units (JBRs) holding monopoly on applied research, and higher education trying to merge within its structures education and research both basic and applied. The interrelationships between economy and science were substantially limited, and lack of competition in the field of research had not been eliminated by adequate institutional measures. Drastic withdrawal of the state from financing R&D section at the end of 1980s and 1990s was caused not only by adopted liberal doctrine of transformation but also by deep recession of Polish economy at the beginning of 1990s. Against the background of general targets of economy transformation process, the objectives of Polish research system transformation were determined as follows (White Book. Poland - European Union: Science and Technology, 1996):

- changing central research control system into determining science policy due to co-operation of the government and scientific circles,
- differentiating the methods of financing basic and applied research,
- introducing competitiveness mechanisms for allocation of budget resources through research project competitions,

- discontinuance of obligatory sending of scientific research findings to economic units not interested in them,
- abolition of the rigid division into three above mentioned sections.

Of course, not all of the objectives were attained to the same extent. (Wasilewski, Kwiatkowski, Kozlowski, 1997). The first three arose from the general system change. Unfortunately, except liquidation of most branch institutes and research-and-development units, the last objective has not been attained yet. (State of JBRs - report, 1996). However companies' approach to innovations has substantially changed. On the one hand they realised more and more the necessity of introducing innovations in order to intensify their own competitiveness. On the other hand the innovation provision was more and more difficult. R&D expenses (GERD) in GNP dropped from 1,4% in 1987 to 0,81% in 1992, 0,70% in 1995, 0,72% in 1996 and 1997. In 1998 they dropped again, which places Poland in the far end of the European list, where they are from 2.5% to 3.0%. Comparisons show that Polish GERD in the middle of 1990s (adjusted by purchasing power parity - PPP - rate according to purchasing power of the currency presented in US dollars) was comparable to Danish and Norwegian GERD. Domestic R&D expenses in Poland were 1.4% GERD of all European Union countries, altogether 0.4% of all OECD countries. (Wasilewski, Kwiatkowski, Kozlowski, op.cit.). Nonetheless sources of funds for R&D haven't changed. Government funds continues to be the main source and companies have not developed in this area much. More detailed data is presented in Table 1.

As the data in Table 1 shows, innovative activity of Poland is rather at the bottom of the list. The relationship between economy and science in the area of innovation projects financing is still unsatisfactory, especially in respect to GERD part in GNP or financial resources per one researcher. It is the worst among European countries even as compared to relatively less economically developed countries (Spain, Portugal, Czech Republic). As statistical data shows the best structure of expenditure relationship concerning R&D in GNP Poland had in 1975 (2.02% which was 10.6 billion PLN). To stop technological gap from developing, that part should decrease, which means that in 1993 R&D expenses should have been 11.2 billion PLN (prices from 1990), so it should have been 175% of the actually covered. During the next years the gap of course widened. If we made such an assumption we could guess that technological innovations would cause economic growth. However current funds shortage concerning R&D section suggests that the economic growth in Poland is due to ownership transformation process releasing effectiveness motivations and as well as to partly realised structural changes, rather than to growth in technological inventiveness. When we connect financing analysis for R&D section with staff analysis making realisation of this section targets possible, we can notice that research expenses in Poland are low as compared with other countries, (GERD in terms of one researcher). That can support the thesis that the danger of exploiting Polish researchers is increasing (the fact that has been observed for a long time e.g. in the activity of foreign consulting companies in Poland).

Table 1: Incomes and technological efficiency in OECD countries in 1995 and 1996

Country	Indicators of technological activity								
	GNP p/c as the % of average GNP in OECD in 1996	R&D expenses altogether as % of GNP	Resear-chers per 10000 work force	Govern-ment funds for R&D as % of GNP	GERD per 1 resear-cher in US dollars PPP	Governme nt funds for R&D as % of the whole R&D	Higher schools R&D expens. as % of GNP	Comp. R&D expens. as % of the whole R&D of comp.	Tech. inten-sity** (inven-tive-ness)
USA	138	2.6	74	0.9	-	34.6	0.4	2.1	10.4
Norway	129	1.7	73	0.8	-	43.5	0.5	1.4	-
Switzerland	123	2.7	46	0.8	-	28.4	0.7	2.2	-
Iceland	117	1.5	72	0.9	-	62.9	0.4	0.8	-
Japan	116	2.8	83	0.6	-	20.9	0.6	2.2	10.6
Denmark	112	1.8	57	0.7	130.651	39.2	0.5	1.7	1.6
Belgium	110	1.6	53	0.5	-	26.4	0.4	1.4	1.8
Canada	109	1.7	53	0.6	-	33.7	0.4	1.4	3.3
Austria	109	1.5	34	0.8	181.038	47.6	0.5	1.1	1.9
France	106	2.3	60	1.0	-	42.3	0.4	1.9	2.7
Australia	105	1.6	64	0.8	177.449	47.5	0.4	0.9	-
Germany	105	2.3	58	0.8	-	37.0	0.4	1.9	5.0
Holland	103	2.0	46	0.9	158.736	42.1	0.6	1.3	3.5
Italy	100	1.1	33	0.5	-	46.2	0.3	1.8	1.0
Great Britain	98	2.1	52	0.7	-	33.3	0.4	1.8	3.2
Sweden	97	3.6	68	1.0	152.799	33.0	0.8	3.9	5.3
Ireland	96	1.4	59	0.3	92.718	22.0	0.3	1.4	1.0
Finland	94	2.3	61	0.9	115.194	35.1	0.5	2.2	2.7
New Zealand	87	1.0	35	0.6	-	52.3	0.3	0.3	-
Spain	77	0.9	30	0.4	90.264	43.6	0.3	0.5	0.2
Korea	69	2.7	48	-	-	19.0	0.2	2.3	0.7
Portugal	65	0.6	24	0.4	92.166	65.2	0.2	0.2	0.0
Greece	63	0.5	20	0.2	-	46.9	0.2	0.2	-
Czech Rep.	46	1.2	23	0.4	96.020	35.5	0.1	0.9	-
Mexico	39	0.3	6	0.2	-	66.2	0.1	0.0	0.0
Poland	34	0.7 0.5*	29	0.26	35.667	64.7	0.2	0.4	0.7
Hungary	34	0.8	26	0.4	57.367	47.9	0.2	0.4	0.7
Turkey	31	0.4	7	0.2	-	64.5	0.3	0.1	-

Source: OECD (1998)

* it dropped to 0.5% of GNP in Poland in 1996

** Technological intensity (inventiveness) is the number of applications for a patent submitted by residents and their impact on the number of quotations in relation to R&D expenses as GERD in US dollars PPP in terms of 10000

At the same time it is surprising that there are no extensive foreign capital investments in development of laboratories, research-and-development units, and that this capital does not do relatively cheaper research. Although Poland has a comparatively large number of researchers (when we compare it to Spain or other countries from the end of the list) but very low expenditure per researcher. Spain spends 2.5 times as much and Denmark with 4 times less number of researchers 12% more. The cheapest are research costs in higher education section (per researcher). For instance in 1994 they were (in millions of US dollars PPP) 15.200 in higher education section, 53.700 in government section and 68.100 in business section. So the relation looks like: 1 : 1.5 : 2.5. The lower GNP in a given country the bigger differentiation. Poor technological intensity in Poland contradicts however other indicators, namely self-sufficiency as far as patents are concerned and dependency. In 1995 dependency indicator was 5.97 while in Germany 1.81, but in Czech Republic 22.48 and Ireland 49.61. Patent self-sufficiency indicator was in Poland in 1995 0.14, but in Spain 0.04. Despite quite good statistical data in this section they cannot be the argument for general concluding because in the EU these indicators are determined the general policy of integration and co-operation. They are however a positive quality in the process of Poland's integration with the EU. The accessory data concerning R&D financing is presented in Table 2.

Table 2: Main financial resources participation for R&D section in Poland in 1995 and 1996

Specification	Resources participation in %	
	1995	1996
All financial resources for R&D activity including:	100.0	100.0
• state budget	60.2	57.8
• companies resources	24.1	28.8
• own resources of research units	11.9	10.1
• foreign resources	1.7	1.5
• rest	2.1	1.8

Source: Yearbook (1998)

Budgeting of R&D section in 1996 was much lower than in 1995 because the very R&D units turned up to be more financially efficient. However, the data seems to be understated. The more because according to OECD estimation also in 1995 in the structure of R&D financial resources the most important part were budget subventions (64.4%).

More interesting seems to be the determination of the relation between R&D financing activity from the commercial and budget sides (Table 3).

Table 3: Companies expenses for R&D   from commercial resources in relation to budget R&D expenses

No.	Country	Commercial expenses in relation to budget expenses %
1	Japan	315
2	Ireland	210
3	Belgium	207
4	Sweden	180
5	Germany	162
6	Great Britain	161
7	Finland	142
8	Denmark	132
9	Holland	110
10	Italy	108
11	Austria	105
12	France	103
13	Canada	100
14	Norway	91
15	Spain	87
16	Austria	75
17	Turkey	53
18	New Zealand	50
19	POLAND	50
20	Greece	44
21	Portugal	33

Source: OECD (1998)

The value of commercial resources and their contribution to R&D financing is an important indicator of research activity of economic units. It can be a direct sign of their initiative in innovations and dissemination of knowledge through involvement in R&D. It is an especially important indicator in economies undergoing transformation, allowing also to assess the direction of industry reconstruction. Poland is in the end of the table ranking OECD countries according to commercial financing. It dominates in the participation of budget expenses in financing the whole of the GERD expenses.

As to structure of financing by R&D different areas it is similar to the previous years when budget resources were transferred mainly to government sector (Polish Academy of Sciences outposts and research-and-development units) - 43.1% and higher education - 31.2%. Only 25.7% of budget resources were allocated in R&D activity of business sector. As much as 73.7% of financial resources obtained by companies were allocated by them in R&D activity, which can be appreciated. However low investments in higher education from business resources can be seen as poor relationships between higher education and industry. Moreover relatively small expenditure for higher education sector - 20.7% doesn't seem to be a good tendency because it means poor possibilities of carrying out researches for numerous highly-qualified scientific staff in academies basing on such narrow resources. Narrow resources appropriated to higher education mean the limitations in number of students in state academies in successive years. Fortunately this gap is partly filled by private education. From the point of view of financial analysis it means that society finance education in Poland doubly - through tax and additionally by financing education in private schools. There has been no development of legal regulations enabling avoidance of double financial burden for people acting in favour of innovativeness growth. Allowances for education of children in private higher schools introduced in 1997 are embarrassingly low (monthly tuition fee).

Taking into account the expenditure structure according to research types with distinguishing basic and applied researches and developmental works, their participation in general R&D expenses in 1995 was : 36.1% on basic researches, 27.0% on applied researches and 36.9% on developmental researches generally. In 1996 these values changed slightly in favour of developmental researches and were : 33.1%, 28.9% and 38.0%. It is said that such a structure is not a positive

Table 4: R&D potential in Poland statistics for 1989-1996

Specification	1989	1990	1991	1992	1993	1994	1995	1996
GERD%	0.90	0.96	0.81	0.81	0.86	0.80	0.70	0.72***
R&D participation in budget %	3.6	2.5	2.5	1.9	1.8	1.7	1.6	1.4
scientific workers number in thousands - altogether including:	66.2	65.1	65.2	63.2	62.7	64.0	64.9	
- in higher schools (full time)	50.5	50.0	51.3	50.7	50.8	52.1	52.9	54.4
- in Polish Academy of Sciences	4.6	4.4	4.4	4.0	3.9	4.0	4.1	3.9
- in research-and-development units	10.9	10.5	9.3	8.4	7.8	7.9	7.9	7.3
Higher schools number (including private)	92	96	117	124	140	160	180	213, 246*
Number of Polish Academy of Sciences institutes	81	77	75	81	82	81	81	82**
Number of research-and-development units	297	260	296	252	310	273	252	255
Scientific equipment value (current prices in millions PLN)	-	359	357	299	370	438	801	-
Scientific and research equipment consumption degree in science and R&D (%)	-	66.5	75.7	80.3	74.4	65.5	73.2	-

Source: Yearbooks (1990, 1996, 1998)

* number of higher schools in 1997

** number of scientific institutions of Polish Academy of Sciences, institutes - 54

*** GERD participation in budget in 1997 was 0.72% in GNP, including 0.5% from budget resources one because it means that basic and applied researches results can be further developed for use in companies and market only to small extent. It should also be pointed out that Polish system of financing and classification is not in accordance with the EU and OECD models. (TPB Manual, OECD 1990). Reasons are to be perturbations in relationships between basic and applied researches and developmental works caused by perturbations in research planning and innovation model maintaining, or it comes to wrong proportions between particular types of research within particular disciplines and research fields.

One of the leading innovative resources is highly-qualified staff. According to our own researches carried out in 1996-1998 and to data published by Central Bureau for Statistics (GUS) and OECD, we express an opinion that because of the general lowering of the funds allocated to financing R&D section in the budget and GNP, the number of people employed in science and R&D activity also was undergoing continuous lowering through the years 1989-1996 and this tendency is still present. (Okon-Horodynska 1998). The exception are higher school staff, which is caused by rapid growth in the number of private higher schools. Staff potential and equipment are presented in the following table (Table 4).

On the base of researches carried out in Poland to have negative impression concerning efficiency of making innovations. In analyses of the process it was taken into account the group of scientific and research workers of conceptual character, making new products and developing new technological processes, as well as the group of technicians (and the like) who implement conceptions under the surveillance of scientists. The other staff consists of administration workers and workmen. Taking into account the number of the employed in particular groups in general level of employment in R&D section 59% are scientific workers, 24.3% are technicians and the like and 16.7% the other staff. Although higher education sector is still dominating, especially if taking into account scientific and research staff it turns out that approx. 31% of time the R&D section workers didn't spend on activity connected with R&D work. It means that the rate of utilisation of the highest qualifications is low. It is connected also with research conditions of poor quality, the scientific and research equipment consumption degree is very high, in many cases scientific workers have to create research conditions on their own. Thus they look for income resources. We know from practice that in recent years as the result of increasing demand for didactic services, large number of R&D staff is involved in extra didactic work in order to improve standard of living. In many cases there is just no time for researches.

Continuing our considerations on the tendency to decrease the level of employment in R&D section, which is enhanced by negative selection of staff (the best researchers leave the section and direct their activity to industry or abroad), we point out that if there actually is such a tendency than the factors behind it cause also a serious weakening of R&D section. This "weakening" is difficult to "make up" even in the period of many years. We can conclude that if the above mentioned process implicates lowering the R&D innovative activity measured in the number of granted and exploited patents (Table 5).

Table 5: Inventiveness characteristic in Poland in 1990-1997

Specification	1990	1991	1992	1993	1994	1995	1996	1997
Domestic patents								
Exploited	4105	3389	2896	2658	2676	2595	2411	2399
Granted	3242	3418	3443	2461	1825	1619	1405	1179
Utility model								
Applied for	2578	-	-	-	-	2119	1800	1589
Granted protections	1694	-	-	-	-	992	1036	885
Foreign patents								
Exploited	1316	1089	1346	1014	1327	1265	1297	1245
Granted	405	371	409	480	735	989	1160	1151

Source: Yearbooks (1996, 1998)

Aggregate number of innovations measured in number of domestic patents exploited in 1997 decreased by 63.6% in relation to 1990, and by 41.5% decreased the number of patent applications. The number of granted protections for utility models decreased as well. On the other hand, however, at the same time increased the activity of foreign inventors in Poland, the number of granted to foreign inventors patents increased by 184%. The largest number of domestic patents in 1995 concerned technology sector - industrial processes and transport (567 inventions, i.e. approx. 21.9% of the claimed inventions). Next place took chemistry and metallurgy (554 inventions, i.e. 21.3%). The least inventiveness activity characterised textiles and paper sectors (34 inventions, i.e. 1.3% of the clamed inventions). Despite weaker activity of Polish R&D section, dependency and independence indicators as it was already said cannot be estimated negatively in relation to other the EU countries. Next disadvantageous fact for activity of the Polish R&D section concerns modernisation scale and production of new products belonging to the aggregate amount of sold production. Indicated indexes are many times lower than in other the EU countries where the participation of new and modified products in the volume of sales amounts to 50%, whereas in Poland this participation was many times lower already in 1980s (6-7%), showed bearish tendency, especially on the early stages of system transformation, i.e. 3.0 - 3.4 % in 1990-1992. Although in the following years of transformation we saw small, 2-3% increase in indexes, it is still very low innovative activity. Relatively highest advance in modification and introduction of new products can be observed in the most involved in privatisation process industries such as electric engineering companies, especially those producing means of transport, paper, wood, glass-making and food industries.

Table 6: Participation of new and modified products in sales of the particular industries (%)

Production branches	1989	1991	1992	1993	1994	1995	1996	1997
Altogether including:	**5.3**	**3.3**	**3.4**	**4.8**	**6.9**	**5.5**	**8.4**	**7.9**
METALLURGICAL INDUSTRY PRODUCTS	**1.1**	**0.3**	**0.5**	**0.7**	**3.1**	**1.1**	**2.1**	**0.7**
• iron	1.8	0.3	0.5	0.9	0.9	1.0	3.1	1.1
• non-ferrous metals	0.2	0.2	0.4	0.3	7.0	1.2	0.2	1.2
ELECTRO-ENGINEERING INDUSTRY PRODUCTS	**13.0**	**8.1**	**9.5**	**14.8**	**15.9**	**16.8**	**17.5**	**18.2**
• metal	10.7	2.2	5.2	5.1	7.2	5.6	14.0	6.9
• machines and equipment	8.6	4.2	9.1	11.3	12.9	10.8	21.5	22.2
• special machines and equipment	29.1	9.0	9.0	5.6	6.9	10.5	12.6	5.0
• transport equipment	10.7	19.4	10.8	22.7	17.8	32.6	21.5	24.6
• engineering products and electronic industry products	18.9	6.6	13.5	12.0	14.9	10.8	11.2	16.1
CHEMICAL INDUSTRY	**6.0**	**2.7**	**5.4**	**6.3**	**4.2**	**3.6**	**7.3**	**5.3**
MINERAL INDUSTRY	**4.0**	**1.2**	**1.4**	**1.8**	**1.9**	**2.1**	**7.9**	**3.4**
• construction materials	5.4	1.2	1.4	0.9	1.8	0.9	8.2	3.0
• glass products	0.9	1.6	1.8	3.6	0.7	6.0	2.9	4.2
• ceramic products	1.8	0.2	0.8	5.1	5.9	3.5	16.2	4.7
WOOD AND PAPER INDUSTRY	**3.5**	**1.7**	**3.6**	**3.4**	**5.5**	**12.9**	**16.7**	**11.2**
• wood products	5.1	2.4	3.0	4.4	8.3	-	-	-
• paper products	0.3	0.6	4.9	1.2	0.1	-	-	-
LIGHT INDUSTRY	**2.9**	**0.9**	**1.5**	**1.9**	**3.2**	**1.3**	**2.5**	**2.3**
• textiles	3.6	1.3	1.8	2.8	3.0	1.7	2.8	3.6
• clothing	1.3	0.6	1.0	0.4	0.3	0.6	1.0	1.0
• leather and fur	2.1	0.3	1.2	1.7	1.3	1.0	4.0	1.4
FOOD INDUSTRY	**1.6**	**0.8**	**1.2**	**1.3**	**2.0**	**1.6**	**3.8**	**4.1**

Source: Yearbooks (1990, 1996, 1998)

For evaluating the inventiveness level of economy the important information is the range of production processes automation, alternative systems used in this area and access and utilisation of electronic media.

Table 7: Production processes automation means and electronic media in industry in 1990-1997

Years	Production processes automation means - in pieces						Electronic media - companies owning	
	Production lines		Processing centres	Industrial robots and manipulators		Computers controlling and regulating technological processes	Local computer networks (LAN)	Access to external computer networks
	Automatic	Computer controlled		Altogether	Robots			
1990	2399	309	566	378	-	2743	-	-
1991	3245	571	670	492	-	4477	-	-
1992	3220	729	654	523	-	4596	-	-
1993	3432	887	744	545	-	5579	-	-
1994	3644	1265	571	594	-	6578	-	-
1995	3999	1568	793	615	342	7909	-	-
1996	4666	2148	1066	744	377	9122	2838	576
1997	5132	2500	1158	950	422	10561	3311	993

Source: Yearbooks (1994, 1996, 1998)

Most robots are used in the production of vehicles, metals, in computer controlled processes concerning mainly gas, electricity and water supplying, as well as machine and equipment production. The above mentioned sections, beverage and food production and publishing activity have access to external electronic media.

The R&D section activity analysis has to take into account the range of licences exploiting as it is a very important transfer canal for foreign technology. Also in this area there has been a significant decrease. Relatively largest number of licences were used in Poland at the end of 1970s and at the beginning of 1980s -it amounted to 329 in 1980. In the following years the number of licences decreased to 119 in 1985 and only 69 in 1988. In the transformation period number of active licences decreased even more to 50 in 1990 and only 38 in 1994. Production obtained due to licences was in 1994 only 2.2% of the industry sales and 5.7% of export production.

3 Conclusion

Generally it can be concluded that R&D section activity in Poland during transformation period shows bearish tendency. However some facts implicating to great extent this situation should be taken into account. Before 1989 large part of innovative abilities were characterised by low efficiency, the funds allocated to innovations were used ineffectively, R&D section staff gained little commercial success and most licences were used ineffectively and in a small way.

Observations made in recent years in the area of R&D section activity allow us to state that although innovative funds are too small, the efficiency of their utilisation is significantly improved. Despite some positive sides the general assessment of R&D section activity is negative. First of all there is lack of comprehensive studies in this area. The researchers in this field, representing few domestic centres (including the author) pointed out, on the base of fragmentary studies, that it is possible to grasp the following regularities:

1. It is possible to observe smaller and smaller number large R&D programs of fundamental significance for economy, accordingly decreases the number of new technologies of high importance for Polish economy and R&D section development.
2. Innovations characterised by highest importance for Polish economy are on a regular base introduced to big privatised companies controlled by foreign investors. Technologies are, however, developed and applied in the first place abroad.
3. In the former economic system the participation of small companies in R&D activity area was negligible. However in recent years the dynamic growth of R&D expenses is really significant, but their participation in expenses on R&D activity in industry is still relatively small, much too small in comparison with participation in production or employment. In small companies there is a tendency to "small" innovations and more important large R&D programs just generally do not concern them.
4. Big companies decrease their innovative activity, especially in long-standing R&D projects involving huge expenses, though in general structure they spend more on innovations then small companies. Perception of technological innovations by companies is defined in a slightly different way than it was made in the introductory part of this article. In the past large research programs were almost completely financed by the state, now companies have to finance themselves through searching for free financial resources which are more and more difficult to find. Thus they use substitutes basing on small improvements demanding far smaller funds and providing quick reimbursement of expenses. Most often new technologies are introduced through equipment and machines purchases made abroad.
5. The independent R&D units were reduced through financial and legal restrictions at the beginning of 1990s. It resulted in disintegration of many highly qualified and efficient research teams and their reconstruction will demand new institutional regulations and will take a lot of time. The other centres and R&D units concentrate their activity on small projects demanding small investments and short time. For most of them this activity acquired productive or service characteristic rather than actually researching or scientific. Therefore working there highly qualified staff didn't make full use of their qualifications. The exception here are the embryos of technological parks or accredited with Foundation for Polish Science centres of technology transfer (CTT).
6. Contacts between large companies and R&D units show bearish tendencies (it concerns 60% of existing research-and-development units), which is the result of financial lacks affecting technological innovations. In the relationships structure large companies are stronger, these alliances are remnants of pre-transformation period. Small companies rarely use the services provided by

R&D units. Activity of bridge institutions, mainly CTT or FEMIRC (Fellow Members to the Innovation Relay Centres in CEECs) in the area of information exchange, conferences organising, contracts scientists, businessmen and investors meetings, did not soften the problem of R&D financing.

7. People giving up or forced to give up their job in R&D units have very high qualifications and it is not difficult for them to find a job. In practice, there is no unemployment among highly qualified staff to which belong R&D section specialists. However they do not work in places where technological innovations should be made.

8. When we think about possibility of creating "knowledge society" it is an important issue for Polish scientific and economic circles to eliminate the generation gap in the area of science, to increase highly qualified staff for science, education and highly developed countries. As regards researches financing and scientific policy Poland diverges from European standards. Harmonisation process in this area will demand increasing of real science expenses. Important issue for long-standing policy of the state is balancing of disproportions between regional research and technologies backgrounds.

9. Some awareness is demanded in Poland. It concerns the direct activity in innovative development meaning introducing to time balance and expenditure of work for creating force in organisations and programs, projects expressing the interest of group of innovators capable of making pressure on realisation of those interests. Therefore this activity is bipolar. On the one hand adequate tools for making pressure should be created, on the other perfect knowledge should be acquired which would allowed correct formulation of research projects, conceptions and methodology of researches, as well as carrying out such research activities that the obtained knowledge products could find buyers in European market.

References

Biała Księga (1996) Poland – European Union: Science and technology. Wierzbicki A (ed.) URM, Warsaw (only in polish available: Polska – Unia Europejska: Nauka i Technologia)

Dobiegała-Korona B (1996) Assessment of the state of the innovation of the polish enterprises. Gospodarka Narodowa, Nr 8-9 (only in polish available: Ocena poziomu innowacyjności polskich przedsiębiorstw)

Felbur S, Czyżowska Z, (1995) Detemination of the financing of science and the budget increase. Ekonomista , Nr 4 (only in polish available: Nakłady na finansowanie nauki a wzrost gospodarczy)

Granstrand O, Sjölander S (1996) Managing Innovation in Multi-Technology Corporations. In: Dodgson M, Rothwell R (eds.) The Handbook of Industrial Innovation. Cheltenham, UK, Brookfield, US, E.Elgar

Krajewski S (1997) Innovations in Large Enterprises During Transformation Period. EAEPE, Athens

Okoń-Horodyńska E (1998) National innovation system in Poland. Katowice AE OECD (1998), Science and Industry Outlook, Paris (only in polish available: Narodowy System Innowacji w Polsce)

Patel P, Pavitt K (1991) Large Firms in the Production of the World's Technology: An Important Case of Non-Globalisation. Journal of International Business Studies, Vol.22, No. 1

Pavitt K (1996) Key Characteristic of Large Innovation Firms. Cheltenham, UK, Brookfield, US, E. Elgar

Business report JBR (1996) Centre of cooperation with the industry, Warsaw (only in polish available: Raport o stanie. Centrum Kooperacji Przemysłowej)

Yearbook (1984) GUS, Warsaw

Yearbook (1989) GUS, Warsaw

Yearbook (1990) GUS, Warsaw

Yearbook (1994) GUS, Warsaw

Yearbook (1996) GUS, Warsaw

Yearbook (1998) GUS, Warsaw

TBP Manual OECD (1990) The Measurement of Scientific and Technological Activities. Proposed Standard Method of Coupling and Interpreting Technology Balance of Payments Data, Paris

Wasilewski L, Kwiatkowski S, Kozłowski J (1997) The development of science and technology. Poland against the background of Europe. OPI Warsaw (only in polish available: Nauka i technologia dla rozwoju. Polska na tle Europy)

Russian Way to Information Society and New Understanding of Information Security (Statement)

Dimitri V. Efremenko

The social debate on the problems of the information society has a specific feature in Russia and Eastern Europe. This feature is related to the dramatic social and political changes in the countries of the Central and Eastern Europe within the last ten years: downfall of the communist regimes, formation and consolidation of democracy, transition from planned system of economy to the free market. These social and political changes are the focus of attention up to now, while the transition to the information society is considered as an concomitant or background process. However, there is good reason to believe that the anticommunist revolutions in Eastern Europe and Russia were the consequences of the first steps towards information society, which is incompatible with the totalitarian one-party system.

The concept of the information society is controversial. There are various interpretations of it. I suppose that the example of the former Soviet Union is an additional argument for interpretation of information society as a post-industrial one. The formation and strengthening of the totalitarian bolshevist regime in Russia was connected with forced industrialisation based on the constraint and centralised planning. Communist ideology however and, in particular, its understanding of the technological progress is one of ideological movements of the industrial era (the source of this ideology goes back to the West-European industrial revolution of the 18th-19th century). Within the framework of communist ideology the total control and planning are considered as means to achieve the social ideal including its technological dimension.

The basic attribute of the information society is the structural and political pluralism. Introduction of the information technology, even the simple spreading of technical means promoting the increasing amount and qualitative diversity of information undermine the total state or party control of information. Accumulation of these changes inevitably leads to disintegration of the totalitarian system.

As is known, all totalitarian regimes of the 20th century used such technological achievements as broadcasting and later television for the purpose of total ideological propaganda aimed at the social and political strengthening of their power. However, the wide spreading of the short-wave radioreceivers had negative consequences for the information monopoly because thousands of people obtained the opportunity to listen to "hostile voices". The development of the computer printing and copying had later similar impact: access to non-censured printed production (so called "samisdat" (self-publishing) that became the germ of free press in present-day Russia) and its distribution were facilitated.

The intensive introduction of the information and communication technology (ICT) and exponential growth of the number of Internet users began only in postcommunistic Russia. But most of Russian Internet users remember the recent past of their country very well. They are aware of the fact that Internet and other ICT are incompatible with the ideological control and information monopoly.

Thus the ICT as postindustrial technologies are not neutral to social and political system. Even the limited introduction and application of the ICT produce the elements of the structural and political pluralism, which result in erosion of the ideological monolith. The introduction of ICT is, nevertheless, inevitable because the premeditated braking of the development of post-industrial technology and the autarchy in the preserve of industrialism and total planning bring to a faster failure of totalitarian regime.

Now other problems of the transition to information society are topical for Russia. Russia as well as other countries of Central and Eastern Europe takes a next step toward the information society. At the same time the general economical, social and political transformation takes place in the mentioned countries. Under the circumstances many people, experts and laymen, concentrate their attention on the problems of the information security and long-term impacts of the ICT development.

Information security is a matter for discussion with participation of scientists, businessmen, developers of software products, users of information webs, government functionaries etc. Information security is also one of tasks for some state institutions of Russian Federation such as the State Committee for Communication and Informatization, Federal Agency for Governmental Communication and Information, Security Council under the President of Russian Federation.

It should be noted that many Russian experts and especially state officers have understood the information security as some sort of the "state security" in old Soviet sense. However it is clear today that information security should not be understood as "state security in the face of threats generated by information society" but, on the contrary, as information society security in the face of inner or external threats. Given this understanding, the state itself can guarantee the security of information society.

When developing an adequate concept of information security one should take into account the nature of new challenges and conflicts related to the transition to information society. First of all it is a question of the risk of a decision making in the information society. The decision-maker has to deal with an ever increasing amount of information, which at a certain stage no more simplifies but complicates a choice of an optimal solution. Today the ocean of information which appears today is similar to those natural elements which plunged the prehistoric man into tremor. At the same time, the difficulty that faces us today is one of searching for information we need for decision making. In the case of ICT the time for decision making is often reduced up to minimum, and in many cases this function passes from man to the electronic systems.

The chances and risks of transition to information society are not reduced only to difficulties of the decision making process under the exponential growth of amount of information and ever developing technical means of its accumulation and transfer. The structural and political pluralism as the inalienable feature of information society comprises inevitably the potential of the conflict among the various purposes and values. The information security in this sense is the permanent searching for the balance among these social purposes and values. Such balance is attainable only under the conditions of civilised and tolerant dialogue, elimination of the abusing by new technological feasibilities and attempts of undermining the principle of structural and political pluralism "from within". The regulative or protective role of the state and international organisations is very important from point of view of the information security. At the same time a self-regulation, dialogue and participation of actors of information society are the factors of ever increasing importance in the course of a decision making. The technological feasibilities for it are constantly developing.

It should be noted, however, that the urgent problem for Russia (especially after financial crisis of August 1998) is the threat of information inequality: comparatively few people can have access to the Internet and other ICT. The role of the state under these conditions together with the regular function of legal regulating of the information market, security of computer webs, protection of state secrets, copyright and intellectual property, struggle against information vandalism etc. should include measures aimed at the overcoming of the increasing information inequality.

The informatization process in Russia of today is characterized both by high intensity and spontaneity. The urgent and serious problems as well as potential consequences of this process are the subject of argument for experts and laymen. At the same time there are the deficiency and retardation of the coordinated efforts of the state institutions and the public for the analysis and forecasting of consequences of the informatization process. As a matter of fact, we should create in Russia a new system of the complex assessment and forecasting of the technological innovations. The institutionalisation of Technology Assessment could become an important step toward the development of such system in Russian Federation.

III The Application Fields of TA

Information Technology and Organisational Change. The Concept of Technological Practice

Gerd Schienstock

1   Introduction

Research on information economy has been dominated for a long time by technological determinism. However, in the frame of this theory, the concept of technology has not been systematically discussed. Most researchers have confined themselves to the description of technical aspects. In the following, we will argue that, to be able to understand social aspects related to information and communication technologies (ICTs), we have to turn our attention to the various forms in which they are applied. However, technological development itself often creates the preconditions for specific forms of application.

In the following, we will discuss various forms in using ICTs. We will argue that increasing global innovation competition causes new ways of applying modern ICTs to coincide with the introduction of a new organisation logic of production and the development of a new organisation culture. Furthermore, the concept of technological practice will be discussed, as it helps us to understand the development and reproduction of technology application, organisation forms and cultural patterns.

2   Alternative forms of applying information technology

As all other technologies, ICTs can firstly be characterised as tools to work with (Eason 1988). The introduction of modern ICTs then refers to single subprocesses or tasks only. Technology from this perspective is visible and tangible; technology is a machine, a piece of apparatus or a device. The tool approach is based on a materialistic concept of technology. As tools, modern ICTs are intended to support people in their work, the aim is to give the person some advantage in accomplishing the task - to enhance the task performance, for example, or to work more rapidly and more exactly. However, information technology could only perform the tool function, when in the beginning of the 1970s the mini-computer was introduced.

Eason speaks of modern ICTs as hand tools. However, the computer or the microprocessor is not meant to manipulate the material conditions of the product but syntactically manipulate its internal state. Therefore, the computer is characterised as 'brain technology' (Krämer 1989). Modern ICTs are very powerful tools as they can store huge amounts of information and process them within a very short time; they increase organisations' memory to an extent not known before (Wash and Ungson 1991, p61). Sometimes it would not even be possible to conduct specific tasks without ICTs as such a powerful tool. Furthermore, because of the fact that it can duplicate and process all symbolic artefacts and that it can easily be reprogrammed, making it a very flexible tool, the computer is in some way a unique tool.

While on the one hand ICTs can become powerful tools in the hands of the workers to support them in their work they can also be used as control device and automation technology. "The most important impact of new ICTs", according to Soete, "is that they move the border between tacit and codified knowledge. They make it technically possible and economically attractive to codify kinds of knowledge which so far have remained in tacit forms" (1996, p49). The codification of knowledge leads to the automation of tasks within production processes and thereby to the elimination of human labour. To codify tacit knowledge, however, is often not easy, and it is especially difficult and costly when the reality it refers to or is operated upon is changing rapidly. So far, automating human skills has proved to be economically viable only in relation to relative simple tasks, but nowadays attempts are increasingly made also to transfer expert knowledge into data and expert systems.

Ernst and Lundvall argue that the elimination of human labour is only one aspect of the automation process, the codification of tacit knowledge and its transfer into modern ICTs also creates demand for new activities and skills. "The very growth in the amount of information which is made accessible to economic agents increases the demand for skills in selecting and using information intelligently" (1997, p28). Modern ICTs, although they depend on codified knowledge, simultaneously create new demands for tacit knowledge.

Concerning the control aspect, the capacity of modern ICTs to store and transmit information on how a task or activity has been performed is important to mention. Due to modern ICTs, it becomes possible to directly monitor the work process and individual work behaviour and to control work progress. Whenever needed, the management can immediately intervene into the work process and can demand a change in work behaviour. We can identify a change from the control of results to process control. Increasing automation allows the control not only of single workplaces, but also of whole work processes. Here the fact that ICTs are open technologies that can be hooked up, with relative ease, to other technological systems (Mackay 1995) is of importance.

The reflexive character of modern ICTs makes it possible also to use the technology for stimulating innovation and learning processes (Zubroff 1988). Castells argues that the information technology revolution contributes to the immediate transformation of new knowledge into innovation activities. Information technology contributes to the direct feedback between the generation and the application of new knowledge (1997, p32). Through continuous monitoring of their work, made possible by information technology, workers can learn and the new knowledge created will then be used to introduce changes.

Furthermore, ICTs are conceived as organisation technologies or organisation strategies. Here the potential of information technology to integrate all information created in work processes is addressed. Digitalisation makes it possible to systematically integrate work processes conducted by human beings and machines into one technical system by abstracting from the specificity of individual tasks and functions. "The growing convergence of technologies, especially in the area of micro-electronics, offers a great technological potential to integrate or merge functions, processes and divisions (van Tulder and Junne 1988, p82f).

The concept of systemic rationalisation points to the strategic aspect of information technology. Borrowing from Weick's concept of 'organising' (1969) we can speak of 'technisation' as a restructuring strategy (Barley 1994). For example, calculable operations can be taken over by machines and then integrated in process overlapping CIM architectures. The technological integration can also cross the border of companies, as, for example, in JIT systems. Modern ICTs therefore become the core of systemic forms of organisational restructuring; instead of focusing on single workplaces a holistic approach is applied, which means that the business process as a whole and particularly its governance structures are the target of techno-organisational restructuring.

Due to the development of telecommunications integrating the computer with information technology, we can identify a new function. ICTs can be conceived of as media that connect people with each other as well as with machines as increasing amounts of communication within and between companies are technically mediated. Information technology is becoming a more and more important device to facilitate and support the information flow within and between companies. The media perspective is dealing with non-formalised communication in the first place. Here the focus is not upon the capability of the modern ICTs to substitute for tacit knowledge, from this perspective the emphasis is on their potential to reinforce human interaction and interactive learning, as Ernst and Lundvall argue, and on how they can support and mobilise tacit knowledge (1997: 28). E-mails exchanged by researchers sharing the same knowledge and applying the same scientific concepts may contribute to such learning processes particularly if all researchers have access to the same databases.

For modern ICTs as media, the following aspects are particularly important:

- the dramatic increase in the speed of communication, with high volumes of data moving from one location to another at rates unimaginable even a decade ago
- a sharp rise in communication bandwidth, with more information of multiple frequencies travelling at the same time down a common line
- the possibility of combining text, voice, video, data, and/or graphics within a multimedia communication system
- an increasing mobility as, due to the miniaturisation process, information technology is no longer bound to a particular place.

It is important to mention that information technology as a medium helps to bridge time and space. Thus a platform for using ICTs as a collective tool is developing. (Baukrowitz et al. 1994). In the case of a decentralised work group-computing model it will be possible to work jointly on a complex task and directly co-ordinate reciprocal sub-processes. For example, workers, even if they are spatially separated, can jointly develop a new product or a new market strategy.

ICTs are also used as a collective tool when a maintenance problem cannot be solved on the spot by the service technician responsible and he has to ask specialists working in the company for support. In this case, the aim of the system development is not to technically replicate whole work processes but to support co-operation through a collective tool and joint access to information.

The newest trend concerning the media perspective is the integration of company internal digital information systems into a public information structure. Two technological developments are important here: interactive multimedia telecommunication applications, on the one hand, and the use of the Internet for commercial purposes including the development of new services on the other. From the viewpoint of companies, the Internet offers an interesting platform for advertising their products and services. At the same time a great demand for new information and communication services is developing. We can expect that in future interactive multimedia applications will be developed which will support tele-co-operation within companies, co-operation within supplier networks and increasingly communication and co-operation with customers.

The following table lists the various functions of modern ICTs in a schematic way and illustrates the aim related to single functions.

Table 1: Alternative perspectives on information technology.

Metaphor	Function	Aim
ICTs as a tool	support for workers in the work process	increase quality and speed, improve capability to cope with complexity
automation technology	elimination of human labour	costs saving
control device	monitoring and steering the work process	avoid defects, adaptation to environmental changes
feedback mechanism	support learning processes	innovation
organisation technology	integration of tasks, functions and processes	organisational flexibility, transparency,
medium, network technology	creation of technical connections among people and with machines	rapid exchange of information and knowledge

Sampler (1996) points to changes in the way ICTs are used. The price-performance improvements in semiconductor technology, new types of technologies, such as multi-media and the improvement of the usability and power of the software, all these developments, Sampler argues, have shifted the fundamental emphasis away from computation towards communication and co-ordination of activities (ibid. 19). Referring to our analysis we can argue that traditional ICTs were mainly used as tools to support individuals in their work activities, as control devices and as automation technology. Instead, ICTs are nowadays increasingly used as organisation and network technologies; they function as media to interconnect people and as feedback devices to support learning processes.

However, one can criticise Sampler's argumentation as being excessively technology-based. Our argument here is that the new forms of applying modern ICTs are mainly caused to develop by the process of globalisation of markets. Globalisation not only contributes to the stiffening of competition but, at the same time, results in the establishment of new competition criteria. Price, quality and time can be seen as entrance barriers to the global market; economic success, however, depends upon the capability of companies to be first on the market with new products which suit the tastes of their customers. In a situation of global competition, innovativeness and customisation are becoming the key criteria for economic success, particularly as the dynamics of technological progress accelerates and the life cycles of many products are shrinking across a broad spectrum of industries. Global innovation competition not only supports the development of modern ICTs applications as it leads to a more rapid exchange of information and knowledge; it can also be seen as the driving force behind the emergence of a new organisation logic.

3   The new organisation logic

The new organisation logic caused by global innovation competition to emerge and develop together with new application forms of modern ICTs completely reverses the key dimensions of the Fordist production model: horizontal specialisation and increased division of work will be replaced by the integration of tasks, functions and processes while vertical integration and the hierarchical control system will be replaced through loose coupling, decentralisation and autonomy in decision-making.

The strategy of vertical de-integration is mainly associated with downsizing and outsourcing. Big companies reduce their size and become leaner to get the same entrepreneurial dynamism, innovativeness and informalism as small companies have and to be able to react as quickly and flexibly as small companies can. Downsizing often takes place together with outsourcing. In companies nowadays more and more functions are becoming the subject of 'make or buy' decisions. If units cannot compete with offers from outside suppliers, they are at risk of being outsourced. Not only peripheral but even functions that can be considered as core businesses, as, for example, design become legally independent through outsourcing. However, former organisational relationships do not automatically turn into market relationships; instead, outsourced parts are often still linked very closely with and economically controlled by the core company based on long-term exchange agreements.

The establishment of cost and profit centres is another strategy to reduce vertical integration. More autonomy is given to the firm's divisions, while at the same time they become fully responsible for costs and profits. Agreements are signed between headquarters and costs centres in which output, the quality of the product, the time of delivery as well as costs or profits are fixed. The tasks and responsibilities of the headquarters, on the other hand, are reduced quite significantly. According to Hedlund and Rolander, in so-called 'heterarchically' organised companies, responsibility for product groups, functions and specific territories can be decentralised in such a way that many units have a say in the company's decision-making process (1990). Loose coupling and direct negotiation among sub-units particularly concerning transfer prices is typical of a profit centre organisation.

A consequence of the introduction of costs and profit centres is the flattening of hierarchies, as organisation levels between headquarters and the centres become obsolete. But the reduction of the organisational levels of bureaucracy continues further down the hierarchical ladder as companies become more aware of the disadvantages of such a large bureaucracy: slow information exchange and even loss of information, lengthy decision-making processes and a non-innovative business culture particularly. All these consequences of large bureaucracies will become serious problems in an economy where business success very much depends on the innovativeness of companies.

The most important aspect of an organisational strategy of functional integration is the introduction of group work. Working groups often become responsible for self-contained production processes without fixed division of labour among the members of the group. Work groups have to organise themselves, they are expected to continuously optimise the way they organise their work. But, at the same time, by monitoring each other, work groups should also continuously improve their co-operation and information exchange. Learning by doing and learning by interacting are important aspects of group work.

Group work necessarily involves reflexivity. As Lash puts it: "It entails self-reflexivity in that heteronomous monitoring of workers by rules is displaced by self-monitoring. It involves (and entails) 'structural reflexivity' in that the rules and resources (the latter including the means of production) of the shop floor no longer controlling workers, become the object of reflection of agency. That is, agents can reformulate and use such rules and resources in a variety of combinations in order chronically to innovate" (1994, p119). There is a tendency to construct rather than simply adopt roles prescribed by management. The group or team approach is also often applied to organise development and design processes. A new style of innovation management reintegrates research and development with engineering, design, procurement, production and even marketing. The development of new products, quality control, market assessment and price calculation – all these activities are performed within a cross-functional team consisting of members from various departments. The idea of such teams is that all products are designed with manufacturing in mind but also to speed up marketing.

Totally new organisation forms appear in outlines, widely characterised as virtual organisations. They consist of various organisation members working together on a project, physically detached and sometimes even out of mobile workplaces. As soon as the project is completed, the working and co-operation structure is dissolved and the group members form new virtual organisations to work on another problem.

Decentralised, process-oriented organisation forms require new forms of co-ordination. Management can no longer rely on the traditional command and control mechanisms but has to turn to more 'soft' measures, such as creating a company vision and developing an organisation culture, organising platforms to exchange information and to develop long-term strategic plans as well as to strengthen co-operation between the autonomous parts. Here we can speak of discursive co-ordination as a new way of governing business processes (Schienstock 1996).

The new organisation logic, we can conclude, manifests itself in two network structures:

- an intra-organisational network structure which develops between sub-units replacing the bureaucratic governance regime; this increasingly includes global production networks as companies break down their value chains and locate discrete functions in different regions all over the world
- an inter-organisational network structure which develops between companies replacing the market as the traditional mode of co-ordinating exchange.

Modern ICTs make it possible for new organisation forms to develop; therefore, they are often characterised as enablers and facilitators of change (Fulk and deSantis 1995). But new organisation forms, in turn, provide new opportunities for technology design. Neither technology nor organisation is fixed but both change in relation to each other; therefore, the design of information technology and the design of organisation forms are largely becoming the same task (Lucas and Barondi 1994, p9). Communication technologies offer opportunities for manipulating both the communication technologies themselves as well as the organisation contexts in which they are embedded, according to specific aims. They provide, as Fulk and deSantis argue, more than traditional technologies, 'occasions' for structuring the production process according to other drivers of change such as innovativeness, cost-saving, control, or quality (1995, p337).

4   The new network culture

The new network structures require, as Schienstock (1997) argues, a new mentality and a new organisation culture. The traditional Fordist model was based on distrust; therefore, a hierarchical control structure was established. A rigid corporate culture developed which is now the most important obstacle to establishing more flexible organisation forms. The use of modern ICTs as a control device aggravated the problem of bureaucratisation and rigidity. A network structure of enterprises cannot function on the basis of distrust and bureaucratic rigidity. In a network where each company is specialising companies become dependent upon each other. Networks therefore need a cultural dimension; they are not purely instrumental, accidental alliances. That kind of co-operation, it is often argued, can only develop on the basis of trust relations.

According to Castells (1996 p199), however, it is not a new culture in the traditional sense of a system of values which holds the organisation networks based on modern ICTs together. A network consists of a multiplicity of actors and networks are very different, which contradicts the idea of a unifying 'network culture'. There is indeed, as Castells continues to argue, a common cultural code in the diverse workings of the network enterprise. But "it is made of many cultures, many values, many projects, that cross through the minds and inform the strategies of the various participants in the networks, changing at the same pace as the network's members, and following the organisational and cultural transformation of the units of the networks." We can speak of a multifaceted, virtual culture. Although it is virtual, it is still powerful in enforcing organisational decisions and co-operation in the network. But, as Castells stresses, it does not stay; the 'spirit of informationalism' is the culture of creative destruction (1996, p199).

Business culture is part of the techno-organisational change process caused by increasing global competition. Taking the more fluid character of the network culture into account, it seems to be more appropriate to understand culture as a 'toolbox', out of which actors can take cultural elements to underline their arguments and to legitimize their behaviour (Swidler 1986, p277). We can then speak of a 'bargained culture'; as soon as restructuring processes take place within the network a new bargaining process starts. This also means that it becomes impossible to crystallise the position in the network as a cultural code in a particular time and space.

5   The concept of technological practice

Technological determinism has been replaced by the 'social shaping' approach to analyse processes of organisational restructuring within and between companies. Recently, Castells (1996), referring to ICTs as a process to be developed, presented an interesting interpretation of the social shaping approach. He distances himself from the traditional concept insofar as he does not differentiate between the process of technology generation and the process of technology application. Instead, he very much stresses both generation and application of technology as being one and the same process. While applying ICTs users at the same time also develop the technology, users and doers may become the same. What follows, as Castells argues, is a close relationship between the social processes of creating and manipulating symbols and the capacity to produce and distribute goods and services (ibid. 32).

However, Castells faces the same critique as the traditional social shaping approach. He also ignores the fact that technological structures become part of the work situation and subsequently condition or even structure the work activities. "Also, in stressing the social nature of the process of technical design and choice, the notion that 'anything is possible' has taken hold" (Wyatt 1998, p18).

This critique takes up the traditional actor/structure dilemma known in general sociology. In his general 'theory of structuration', Giddens (1979, 1984) has made the attempt to resolve the dualistic grouping of objective conditions and subjective action and interaction and to integrate the actor and structure perspective in one theoretical framework by introducing the concept of 'social practices'. According to him, agents and structures represent a duality; they are not independent sets of phenomena but a dialectic relationship exists between them. Structures are the product of but also a medium for social action; they are not only barriers but also give momentum for social action. When reproducing specific social practices, actors have to draw on existing rules which at the same time structure social practices.

Structures, according to Giddens, do not exist in reality; they only exist as memory traces in human minds. People internalise the rules that are made to manifest themselves only in the instances when they are drawn on in social action and interaction (Walsham 1993, p61). Structures exist insofar as they are produced and reproduced in social practices. Therefore, we can also speak of 'virtual structures.'

Social rules and structures, on the other hand, enable people to act in a competent and context adequate manner. In acting according to the rules human agents not only reproduce existing social practices but they also confirm their institutional character. However, there is always the option of 'doing otherwise'; every social actor controls some resources which give him the power to influence existing practices. Although the disposition of social actors is imprinted to act in a specific way, they always have the option to change social practices; they do not have to reproduce them schematically. Human agents monitor their own conduct and its results in a reflexive manner and if they do not produce the expected results, they can either act differently, changing existing rules, or they have to adjust their goals and expectations. This reflexivity together with the concept of unintended consequences of purposeful action implies "that all action carries within it the seeds of change" (Walsham 1993, p61).

Giddens' general theory has a great disadvantage insofar as he does not pay a great deal of attention to the role of technology in the production and reproduction of social relationships (Wyatt 1998, p20). Modern ICTs, however, are not mere social constructions that can be reduced to a set of social relations. While social relations are undoubtedly present, they do not comprehensively or essentially describe what those technologies are. ICTs possess an objective set of rules and resources that both enhances and constrains the roles of workers within organisations, they have an influence on the social fabric of the organisation (Samples 1998, p10). To conceive of a social order as resting solely on interactions between human beings is to ignore the role that material resources play in shaping and substituting social relationships (Kavangh and Arujo 1997).

In technology research, the concept of 'technological practice' (Parcey 1983) represents an approach that also claims to transcend the duality of structure and process or the 'social shaping of technology approach' and the 'impact analysis' of technology on society. Relying on Parcey's definition, technological practices are seen as having three dimensions: a technological, an organisational and a cultural (1983). The three dimensions are very much intertwined: changes in one dimension also affect the two others, which means that all three dimensions have to be developed at the same time. However, the approach does not deal with the question how people actually use technological artefacts and how artefacts are involved in the creation of social structures and relationships.

Here the technological practice approach has to be developed. At least three aspects can be mentioned as critical for this approach:

1. It needs to be a 'bottom-up' approach, this means that the analysis must focus on the process of designing information systems in different intra- and inter-organisational fields. The analysis must focus on social actors involved in the production, reproduction and change of technological practices. However, to avoid a simply voluntaristic view, the approach has to be extended to the question of how ICTs are involved in the creation of social structures and relationships.
2. We cannot assume that modern ICTs determine organisation forms. Instead, modern ICTs can be characterised as progressively freeing agency from structures; technical and organisational structures both become the objective of reflection of social actors. They can rearrange their work relationships in order to make the production process more effective, flexible or to increase the learning capacity. Therefore, the formation of information economy can be analysed as the production of changing relationships between actors and systems.
3. The approach also includes a new way of conceptualising technology. Instead of dealing with technical dimensions only, the focus should be on the various ways of using modern ICTs. How people actually use ICTs and what functions these perform should become a key research question.

There are two more questions related to the 'technological practice approach'. Taking into account that technological practices are socially embedded we have to analyse in what way they are influenced or formed by their environment. But it is also interesting to analyse, in what way technological practices influence their economic and social environment.

The following picture presents the key aspects of the technological practice framework in a schematic way.

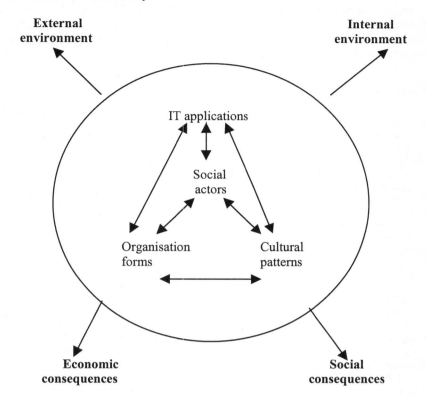

Figure 1. The concept of technological practices

6   Conclusions

Research on the information economy was imprinted by the approach characterised as technological determinism. It is argued that technological development determines changes in the organisation of work. Here we have argued that instead the increasing globalisation of markets and innovation competition associated with it has to be given priority as a causal factor for the renewal of the economic system. This does not mean that technological determinism will be replaced by another variant of determinism, which can be characterised as economic determinism. Global innovation competition, we argue, entails a new logic of applying ICTs and organising work and it also implies a changing business culture; however, such new production logic can be translated into production practices in different ways.

Relying on Dosi's (1982) differentiation between a technological paradigm and various technological trajectories, we here propose a distinction of a new informational production logic or paradigm from its various national or industrial trajectories. Thus we can avoid the purely voluntaristic approach, characteristic of the traditional social shaping approach without neglecting the contribution of social actors to build up information economy.

References

Barley SR (1986) Technology as an Occasion for Structuring: Evidence from Observation of CT Scanners and the Social Order of Radiology Departments. In: Administrative Science Quarterly 31, pp 78-108

Baukrowitz V (1996) Neue Produktionsmethoden mit alten EDV-Konzepten? In: Schmiede R (Hrsg.) Virtuelle Arbeitswelten: Arbeit, Produktion und Subjekt in der Informationsgesellschaft, Berlin, Edition Sigma, pp 49-71

Castells M (1997) The Rise of the Network Society, Maldon Mass./Oxford, Blackwell Publishers

Dosi G (1982) Technological Paradigms and Technological Trajectories. A Suggested Interpretation of the Determinants and Directions of Technical Change. In: Research Policy 11, pp 147-162

Eason K (1988) Information Technology and Organisational Change. London/New York/Philadelphia, Taylor & Francis

Ernst D, Lundvall B-Å (1997) Information Technology in the Learning Economy - Challenges for Developing Countries. DRUID Working Paper No. 97-12 of the Danish Research Unit for Industrial Dynamics, Aalborg

Fulk J, deSantis G (1995) Electronic Communication and Changing Organizational Forms. In: Organization Science, Vol. 6, No 4, pp 337-349

Giddens A (1979) Central Problems in Social Theory: Action, Structure and Contradiction in Social Analysis, London

Kavangh D, Arujo L (1997) Folding and Unfolding Time. Accounting Management and Information Technology 5/2, pp 103-121

Krämer S (1989) Geistes-Technologie. Über syntaktische Maschinen und typographische Schriften. In: Rammert W, Bechmann G (Hrsg.) Technik und Gesellschaft. Jahrbuch 5: Computer, Medien, Gesellschaft, Frankfurt/New York, pp 38-52

Lash S (1994) Reflexivity and its Doubles: Structure Aesthetics, Community. In: Beck U, Giddens A, Lash S (eds.) Reflexive Modernisation. Politics, Tradition and Aesthetics in the Modern Social Order, Cambridge: Polity Press, pp 110-173

Parcey A (1983) The Culture of Technology. Oxford/New York

Sampler J (1996) Exploring the Relationship Between Information Technology and Organisational Structures. In: Earl M (Ed.) Information Management. The Organisational Dimension, Oxford, Oxford University Press, pp 5-22

Schienstock G (1993) Management als sozialer Prozeß. Theoretische Ansätze zur Institutionalisierung. In: Ganter H-D, Schienstock G (Hrsg.) Management aus soziologischer Sicht. Unternehmungsführung, Industrie- und Organisations-soziologie, Wiesbaden, Gabler Verlag, pp 8-46

Schienstock G (1997) The transformation of regional governance: institutional lock-ins and the development of lean production in Baden-Württemberg. In: Whitley R, Kristensen PH (eds.) Governance at work: the social regulation of economic relations in Europe, Oxford, Oxford University Press.

Soete L (1996) Social impacts of the information society - National and community level. In: Finnish Institute of Occupational Health (ed.): Work in the Information Society, Helsinki, pp 45-55

Swidler A (1986) Culture in Action: Symbols and Strategies. In: American Sociological Review, 51, pp 273-286

Walsh J, Ungson G (1991) Organizational Memory. Academy of Management Review, 16, pp 57-91

Walsham G (1993) Interpreting Information Systems in Organisations, Chichester, John Wiley & Sons

Weick K (1969) The Social Psychology of Organizing, Reading Mass, Addison Wesley

Wyatt S (1998) Technology's Arrow. Developing Information Networks for Public Administration in Britain and the United States, Proefschrift, Universitaire press Maastricht

Zuboff S (1988) In the Age of the Smart Machine, New York, Basic Books

Data - Information – Knowledge.
A Trial of a Technological Enlightenment

Klaus Kornwachs

1 Introduction

The main issue of the triad „democracy - participation - technology assessment"
has been focused on the transition from an Information Society to a Knowledge
Society. First of all, this transition can be considered as a transition from one
headline term to another one within the context of socio-technical research
programs.

In order to resolve some confusions it seems to be necessary to look for the
relations between data, information and knowledge. Speaking from a xy-society as
an attempt to state that xy is important and decisive for an economic exchange of
goods and efforts, we could state that we are living in a society, where the variable
xy may have the value xy = capital or even xy = labour or xy = information or xy =
knowledge and so on. Regarding the concrete societal and economic reality, all
this values of the variable xy can be justified easily taking particular aspects into
consideration. Therefore the term Knowledge Society seems to be empty as long
as one is not able to say something about the difference between knowledge and
information. In my opinion, this difference is a very crucial one and it cannot be
explained without technical and philosophical terms.

The relations between knowledge, information, communication and data
processing can be described by our knowledge from technology, i.e. information
and communication technology. But we have no common sense definition of
knowledge – even in philosophy there is lacking an unanimously defined
definition either. Some steps within a scheme modelling an overall relation from
process observation, signals, signs, data, information, up to knowledge are well
known, for some other steps adequate models are missing. Speaking about
Knowledge Society and Knowledge Technology one should be honest enough to
put emphasis on this deficits.

The famous step from data to information can be characterized by the
circumstance that we don't know how the semantic closure is generated that is
used implicitly by any every day user of a data processing device or of a
communication network. This semantic closure is determined by organizational
conditions, by technical facilities, but mainly by mutual understandings and
consent in a formal and substantial way. The Informed Society (this term has been
made known by Minc and Nora 1979) is based upon such understandings and
regulations. But these regulations and consents has been cancelled by new

technological possibilities and new organizational issues have been forced to be negotiated and to be accepted.

It seems to be an important point that knowledge about the inner structure of the sociotechnological systems with respect to their organizational and semantic closures is necessary in order to analyse more deeply how the future societal structures will develop. This knowledge is a contribution to something I want to call „technological enlightenment" according to the term „technologische Aufklärung" put forward by Ropohl (1991).

Talking about Knowledge Society remains empty as long as there is no reasonable model about a distinction between information and knowledge. To be more precise – a model is needed how knowledge is generated by information. Information is a quantifiable, measurable concept, information may be transported, transferred, stored, annihilated, it is bound to a material carrier. But knowledge seems to be bound to a human carrier and it is closely connected with the concept of action. On the one hand knowledge is a kind of an active process (an action), on the other hand we use this term to describe the results of cognitive acts. Knowledge as description can be related to facts, properties and real processes, knowledge as prescription (rules, orders, commands etc.) can be related to the area of ought, obligations, wishes and will or – representatively – to dispositions and abilities.

The concrete meaning of the term „Knowledge Society" remains still unclear.[1] If knowledge is conceived to be the stuff future is made with, then the notion of Knowledge Society is only a metaphor, disguising that we don't know what knowledge is within this context. Obviously we try to materialize or naturalize knowledge when having the conviction that knowledge could be treated as an economic good. This has been done by the sociology in the 80ies with the concepts of message and information - theses trials have not been proved to be very successful and they didn't contribute to an enlarged understanding of the potentials of the information- and communication technologies (ICTs).[2]

It is struggled intensively until now whether it is possible to solve the emphatic[3] concept of knowledge from its anthropological background and to attach cognitive processes consequently to artifacts, machines or systems, presupposed that such kind of processes are necessary conditions for the generation and use of and handing on knowledge by transferring information.

Behind this struggle we may detect philosophically on the one hand the debate about artificial intelligence and the mind-body problem. From the point of view of Philosophy of Technology we may face the problem that the potentiality of producing artifacts is limited with respect to their functionality and certain properties by formal (cf. Gödel, Turing) and physical (thermodynamics, quantum mechanics) restrictions. Nevertheless we have available a universal instrument for producing virtual artifacts by the mean of digitalizing the information and communication technologies and their merging and emerging potential. Such virtual artifacts could be programs (software), forms of organizations,

---

[1] It has been introduced by Stehr (1994).
[2] Needless to say, that the classical economical concepts fails in applying on informations and knowledge (cf. Zerdick et al. 1999).
[3] At least in Western philosophy knowledge is an emphatic concept since Plato.

communication protocols, knowledge bases, expert systems virtual worlds (like in architecture), games, supervision systems etc. This virtual artifacts are not restricted to physical boundaries and we do not know the formal limits yet. Such artifacts determine - in a certain way - autonomously the rules how to deal with them, because they are extremely sensitive against the semantic and organizational closure in which they are produced, used and applied. Therefore there are subject of fearful technology assessments. Nevertheless, artifacts - as material, technical, immaterial, organization or virtual as well - has been always produced driven by interests.

Very often in information and communication technology we are speaking about systems and we are thinking on such artefacts, systems, that contains technological components and organizational closures. It may be a task of technological enlightenment to make clear explicitly the meaning of speaking about systems. System theory is a description tool, no theory about a social or natural ontology. There are properties of systems than can be explained by analogies in terms of physical, technical, formal, organizational and logical disciplines, as well as information theory and computer science. But each system has an author, it remains as a description, not as an existing entity. The other way around, the next task of technological enlightenment should be to examine this terms more precisely, even if they are used or abused for metaphorical purposes like in sociology.[4]

It might be interesting to observe that the multitude of meanings of this terms about data, information and knowledge is strongly decreased, if technological necessities, economical issues and the question of power came into play. Then we will reduce data as the basis for control, information becomes useful or not and knowledge is attached with the Baconian *ipso scientas potestas est*.[5] The most important insight is that not technology, but our technological, economical, social and political will is determining what might be useful, adaptable and adequate to certain structures of power.

2   Cognition, Consciousness and Information[6]

Let us begin with a very simple question: How can I prove that a stone lying in front of me, is not thinking? This question is closely related with the problem of recognizing cognitive processes outside of the own consciousness.[7] The only fact that Descartes could take for granted was the experience that he is himself thinking (*„cogito, ergo sum"*[8]); this was a proof for his own existence.

One is inclined to go so far that if the recognition of the own mental processes is so difficult to achieve, than the positivistic question would be much more

---

[4]   Cf. the debate about Luhmanns "Social Systems" (1991).

[5]   Francis Bacon: Science is power (≠ knowledge is power) cp. Bacon (1597/1864), p. 79.

[6]   Chapter 2 has been adapted and modified from a publication appearing soon (Kornwachs 2000).

[7]   In Philosophy the problem is known as „Erkennung des Fremdpsychischen" (i.e. recognizing other psychological systems) ; cf. Carnap (1967), Orig. in German Carnap (1928).

[8]   In Latin cf. René Descartes: Principia Philosophiae I, 7; in French „Je pense, donc je suis" in: Discours de la Méthode IV, 3).

difficult: how can I recognize that an entity outside of me may be a thinking entity, or in other words, may be performing mental acts, i.e showing consciousness, emitting meaningful information, having knowledge.[9] In philosophy, particularly in analytical philosophy there is a well known tradition of such questions and with the rise of computer and information technology this question has been connected with the problem of artificial intelligence and the simulation or even creation of cognitive processes that are supported or generated by a machine[10].

If one is considering that the notion of Knowledge Society presupposes that there is somewhat receiving and understanding information in order to acquire useful knowledge, the nature of cognition should be clarified as an elementary act constituting this kind of society.

Thus, to interpret a behaviour as a result of a cognitive act presupposes that the observer must have already a pre-knowledge about possible goals, intentions or will of the subject observed. If the subject observed is tried to be described *as* a system (i.e. is modeled with means of system-theoretical description tools) nothing is said about the nature or essence of the subject.

---

[9] I confess to have intermixed thinking, performing cognitive processes, performing mental acts, showing or having consciousness as synonyms. But all these terms are homonyms also. It is a wide spread opinion that thinking seems to be reserved for philosophers and other people. Nevertheless animals are also able to perform simple cognitive acts, e.g. recognizing something. Another prejudice exits - only the having of (self)-consciousness is allowed to be correlated with being able to perform mental acts. Despite of these opinions I do not distinguish here much more precisely than one is able to reflect about it.

[10] The literature is exploding, so only some selections: Churchland 1986, Carrier Machamer 1997.

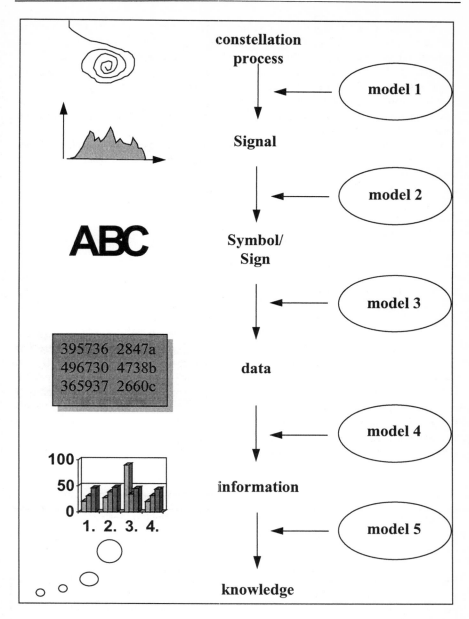

Fig. 1: Multilevel concept for pure perception of constellations or processes, signal processing, symbol writing and reading, data processing, information exchange and knowledge generation and use. Details see text.

The same holds for any entity like a machine or any other device that is believed to show cognitive processes. A cognitive process must have a result. The result should be observable by an outside observer, otherwise no statement about such a process is possible. The Cartesian self-observation of cognitive processes by the subject actually performing it, may provide a high degree of (self-)conviction, but this conviction holds only for the subject itself. If a cognitive process shows some results on the outside surface of the subject performing it, then this must be either a constellation, a signal, a sign, some data, information or a knowledge transfer that give rise to the interpretation, that the subject may have performed an act of understanding something.

In the following subchapters, a multilevel concept (see Fig. 1) is presented that tries to explain how the distinction between pure perception, signal processing, symbol writing and reading, data processing, information exchange and knowledge generation and use of knowledge can be distinguished.[11]

2.1
Constellations

Understanding means an adequate endo- and exo-(re)action on stimuli that come from outside or inside the subject (including completely autonomous selfreactions). Conceptualizing cognition in such a way, cognition means at least the understanding of the outer physical world, i.e. effects, processes, constellations, and situations (see table 1). Effects are mediated by physical states and their dynamics; they can be stimuli by the mean of sensory perception. Processes can be recognized by ensemble of timely ordered effects and here not only sensory perception is necessary but also perception which is mediated by the memory over a time period.. Constellations or situations are perceived by processes and properties, but the recognition of properties presupposes already a certain kind of *a priori* concepts, achieved before by mechanisms that are struggled since Kant. Beside sensory perception and memory the ability of pattern recognition is necessary, and one could say that pattern recognizing is one way to achieve *a priori* concepts in order to recognize what will be a good pattern later.

Table 1: Outer physical world by effects, processes, constellations, and situations.

• **Effects**	= Physical States and theirs Dynamics,	Sensory Perception
• **Processes**	= Ensemble of Effects,	Sensual Perception with Memory,
• **Constellations**	= Processes and Properties,	Sensual Perception with Memory and Pattern Recognition,
• **Situations**	= „Colored Constellations", i.e. Metaproperties,	Sensual Perception with Memory, Pattern Recognition and Semantically Leveling

---

[11] The multilevel concept has been developed during my lecture at the University of Ulm in Summer Term 1994. There are some coincidences with the „Descriptions and effects levels for intelligent system behavior" in Radermacher 1996, cf. Fig. 1, p. 4.

The recognition of situations in the outer world could be considered as a kind of „Colored Constellations", i.e. metaproperties. Sensory perception with memory, pattern recognition and semantically leveling is necessary to perform such an cognitive act. The scheme in Table 1 tries to comprehend this view.

2.2
Signals

It is not far fetched that since the enlightenment the outer world as far as it is recognizable by us, does not show any willingly or intentional issues - one of the great successes of modern science since Galilei and Newton was to eliminate the concept of goal from natural science.

Obviously each mental system is processing signals from the outer environment and considering this fact from an exo-view there is no sufficient condition that could be extracted from the signal itself that it is containing a meaningful input for the receiving subject or entity. This is the reason why the „Mathematical Theory of Communication" by Shannon and Weaver (1948) is not dealing with meaning explicitly.

The interpretation of a change of physical states (Table 2), i.e. to observe a time series, presupposes the concept of an observable variable and this concept is already theory laden. In other words: Even the simplest observation act presupposes a model within the receiver, that encompasses the concept of a concept, the ability of feature extraction and last, but not least - a concept of a time order. Singular events like flashes or bursts need a concept of an event[12] that is epistemological not trivial. If a signal is realizable as a parallel presentation, like a landscape or a picture, a concept of space (two or three dimensional) is required. The mentioned model on this level is not reducible to the model that is required to recognize constellations. To be more precisely: The model 1 (level 1 → 2) that is required here in order to interpret a constellation as a signal is not reducible to the model, required to observe constellations or situations (say model on level (0 → 1). A short scheme in Table 2 will sum up this.

Table 2: Signals

• **Time series**	= Physical States (Variables),	Concept of Concept, Feature Extraction, Concept of Time order
• **Singular Event**	= like flashes	Concept for an event
• **Parallel Presentation**	= landscape	Concept of Space

---

[12] Heidegger (1988, p. 20 ff. engl. 1972, p. 24)) is calling an event (Ereignis, old german: Eräugnis, i.e. the way to observe with eyes), using a wrong german etymology, as an Er-eignis, something, what is happening at a time that is closely related like an eigenvalue in quantum mechanics or eigen-behavior (von Foerster), i.e. it cannot happen beside its appropriate time.

2.3
Symbol and/or signs

The next level (Table 3) can be regarded as the processing of symbols and signs.[13] A signal can be interpreted as a sign or symbol, if there is a pre-knowledge that allows the receiver to refer on that for what the symbol or signs stand. At least at this level semantic considerations are unavoidable. The „meaning" of a sign is on what it can be referred, but this act of being refereed to something is done by the receiver or processor of the sign. Regarding a picture or a painting the observer must be familiar with the situation the picture is representing, otherwise it may possible that the receiver is not aware that the present painted wall could be a picture at all. Knowledge about the world is already necessary to distinguish between sign, representing some objects to which the sign is referring to and signs that indicate that something is happening or something is the case. „Where smoke is, there is fire" means that smoke is an indicating sign for fire. To realize that, a lot of knowledge about the world is already necessary.

The starting point in Shannon's-Weaver theory of communication is the concept of a repertoire of isolated signs. They represent events or decisions or a choice from an n-fold alternative. This n-fold alternative or even a repertoire of signs means an already perfect prepared situation which cannot be interpreted in a naturalistic way: In communication situations the choice of signs from the repertoire is goal oriented, in biology the discussion of finite n - fold alternatives is just a *façon de parler* to apply entropy measures. Again the model, needed to interpret a signal as a sign, is not reducible to the model that is necessary to interpret physical constellations as signals.

On this level it makes sense to conceptualize quantitative measures like Shannon entropy in order to assess the amount of statistically measured information that a signal may contain. Thus a necessary condition can be formulated for the distinction between pure stochastic (noisy) and information containing processes.

Table 3: Symbols and/or signs

• **Picture**	Iconographic	Familiarity with the situation
• **Referring Event**	Where smoke is there is fire	Knowledge about the world
• $Z \in \{\Re\}$	Signs from a repertoire (numbers, letters)	Prepared situation

---

[13] In order to facilitate the interdisciplinary dialog I omit here the traditional distinction between symbols and signs. Within a formalism a symbol stands for the name of an entity that belongs to a certain category or class of entities, like a variable (logical sign). A sign stands for something that is not present or available, like a word. It is used as an instrument in order to refer to something within a communication situation (semiotic sign). Both, symbol and sign, are representing (or referring to) something else that is not identical with them. This common property is sufficient for our present reflections.

2.4
Data

Going to the next level (Table 4), it is trivial that data contain signs and that data are set of signs that can be handled like stored, transferred, ordered, written and deleted. As a technical concept data processing needs the materialization of signs, i.e. a physics of reading and writing. What can be figured out from data by data processing or interpreting them is no more invariant under permutation of the order of data and the selection of data within time order (so called mask). If the data are not explicitly masked they can stored in a time order at least. A mask is not derivable from the data itself without any outer constraint (e.g. maximum or minimum entropy), it represents an exo-view respectively a preparation act done by the receiver or another exo-agent. Coding means here the mapping from one repertoire of signs to another one or the same repertoire. The results of data processing (say with the same algorithm) are not invariant under arbitrary coding procedures. Coding requires in advance a preceding communication act to establish a mutual understanding about the rules of encoding and decoding.

From a point of endo-view data processing is nothing else than to manipulate the order and arrangement of given signs, as J. R. Searle has demonstrated with his example of Chinese Room[14]. The processor does not know anything about the meaningfulness of the data processed or better: in order to perform the procedure, expressible as an algorithm, it is not necessary for it (him?) to know all that. The performance is independent from the reference of the signs that are processed or rearranged.

The usefulness of data processing due to what has been called the semantic closure in chap 1.[15] This expresses the simple precondition that the meaning of the signs, contained within the data, is given by a reference procedure necessarily performed *before* processing it by an algorithm. Mostly the computer user knows the meaning of the input data and what the program is doing with them (for instance: to sum up numbers that stands for money or tons of corn). The interpretation of the result, i.e. whether the output data are meaningful for the user or not, presupposes necessarily (not sufficient of course) that the output data are interpreted under the same referential frame. Therefore the interpretation context of input and output must be the same.[16]

As a consequence computers and even parts of entities with consciousness that perform only data processing do not generate meaning nor do they generate information as it is said sometimes in a popular way. Again we have a scheme like the following.

---

[14] All the debates about the question „What Computers can do or not" are discussing on this level. Searles Chinese Room (cf. Searle 1987) is a modified version of the Turing Test (cf. Turing 1950). Both situations represent a „Gedankenexperiment", whether a pure data rearrangement can produce meaningful information.

[15] This expression has been introduce by Pattee (1977) in terms of cell and molecular biology.

[16] This is not trivial: the semantic closure can be destroyed by separating the input and output with respect to the communication situation of the people involved or the meaning of the output may be simply forgotten after a certain time.

Table 4: Data

• **unmasked**	Time order	storage
• **masked**	Mask	exo situation, prepared
• **coded**	Repertoires	mapping and pre-communication
• **meaningful**	semantic closure	context of data

2.5
Information

After having separated the concepts of constellations, signals, signs and data in order to distinguish several levels, the *façon de parler* of „information processing" seems to be misleading. Information is something that can be understood and therefore with respect to a subject, there is a distinction possible between potential and actual information and a further distinction between information needed for something about something and information from something for somebody. This distinction can be found also in advanced information theoretical models in the realm of neuronal science.

The information concept as it will be developed within the next chapters is a pragmatic one (see Table 5): Within a computer data remains data-type data, the computer as a kind of channel remains stable. In biology where we have processes like growing and dying, rise and decay despite and because of a remarkable stabilization power, there are learning processes on different levels that can be observed as the change of structure and behaviour of the organisms, systems or subjects involved. Therefore a measure for information should be here conceptualized a measure of its effect on the receiver. In order to operationalize this fact, the receiver must became a sender and vice versa.

One can express, according to a proposal of E. U. v. Weizsäcker (1974) the expectations for these changes in terms of novelty (or surprisal) and confirmation (or redundancy). With a given semantic closure it is possible to find out whether an information has been understood or not observing the reactions of and changes within the receiver. From a syntactical point of view information processing and transfer can be described as a (possibly infinitesimal) chain of series - spatial transformation of signs, ordered as data and processed within a semantic closure. But this is only a necessary, not a sufficient relation. Another necessary condition is that there must be an entity that is able to be changed under the influence of information. Physically spoken the change of constraints of a process can cause effects that are comparable with learning or structural change.[17] But a sign or a set of realized signs as data can become only information when they are able to change the receiver. In terms of set theoretically conceptualized probabilistic information theory this would mean that information as an acting entity can be observed as the change of the probability distribution of the expectancies, as the generation of new alternatives within an n-fold alternative and the change of the surface of the affected system as well.[18]

---

[17] Cf. Haken 1988, Gernert 1996, 1985.

[18] Surface means here the separation between inside and outside of an entity described as a system. The behaviour is mediated by the surface and therefore information is affecting the surface; cf. Kornwachs (1996).

Table 5: Information

Series - parallel-presentation	syntactical structure	Reordering processes and transformations
Receiver becomes sender	pragmatics	Expectations: Novelty and Confirmation
Acting entity	changing systems as source and receiver	Conditioning, learning, reacting, structuring
...		

2.6
Knowledge

Knowledge (Table 6), often confounded with information, can be considered as a result of understanding information, that has been integrated in already existing knowledge. I would like to hypothesize that there is no knowledge without already existing knowledge. Thus the first knowledge must have been come into existence without the first information understanding process. This pre-knowledge reminds to the innate ideas[19] or the "Anschauungsformen" by Kant.[20] But to integrate information within a corpus of an already existing knowledge it is necessary to bring information from an exo-knowledge (about the world) to endo-knowledge (how we are perceiving the world). In other words: The transformation of sensory signals to data, the transformation data into information are processes that lead from the outside of the receiving entity to inside. Knowledge is not only referring to the outer world. We are speaking and thinking about world, we also need a reference to the distinction between inside and outside and all this presupposes a surface of the speaking and thinking entity. Together with the requirement that data are physically realized and ordered set of signs, one could say that signs are „living on the surface" of the entity.[21] Knowledge cannot exist disembodied - it presupposes body and the experience of it.

This circumstance is often forgotten by the protagonists of an artificial intelligence program designed in pure algorithmic terms. The integration of information into the realm of already existing knowledge is done by modifying this knowledge, i.e. understanding *is* modifying knowledge. The contrary process is also possible: we are able to forget portions of knowledge without loosing the whole knowledge and its integrative power for new information understanding. This can be considered as a modification, too.

But the most interesting thing is the ability of cognitive systems to inform themselves, i.e. to aggregate, to comprehend, to extrapolate and to re-contextualize portions (chunks) of knowledge. The acts of forgetting,

---

[19] Cf. Platos dialog: Parmenides (128c - 135b), the concept of innated ideas at Descartes: Principa Philosphiae I,10, and Letter to Mersenne from 16.6.1641 (Oeuvres III, p. 383). Chomsky (1966) has used the term of innated ideas when he was trying to explain how children are able to learn the syntax or a language without difficulties wheras the theoretical structure of grammar is very complex.

[20] Cf. I. Kant: Critique of Pure Reason. B 34, B 75.

[21] This metaphoric expression I owe to intensive discussion with Wolfgang Hinderer.

remembering, imagine, and even thought processing are lying on this level. We sum up again like the following:[22]

Table 6: Knowledge

• **Understanding**	Integrating information from exo- into endo-knowledge	New structures and surfaces
• **Modifying**	Restructuring own knowledge	Understanding is modifying
• **Deleting**	Forgetting, „Verdrängen" (supress)	same
• **Generating**	Operating on the own knowledge base	Aggregating, comprehending, gap filling, contextualizing

2.7
Realization and Models 1 - 5

Within the following Table 7 the forgoing discussion is tried to be forged to a certain synopsis. The mentioned models for the steps from processes → signals →symbols → data → information → knowledge are different, they are based on different mother disciplines that cannot be simply reduced to each other. It is clear that knowledge contains information, that information is represented and transported by data as by symbols as by signals as well and that signals are based on observable physical processes. But the other way around it is hard to find relevant scientific disciplines that may answer the question how to achieve the chain from processes → signals →symbols → data → information → knowledge.

---

[22] Cf. also Kornwachs (1999, 1988).

Table 7: Models for the steps from Processes → Signals →Symbols → Data → Information → Knowledge. The rows indicate the entities and levels of the relevant step the respective transforming processes, i.e. what is necessary to interpret the forgoing entity *as* the following one. The relevant models for the transformation n → n+1 are not reducible to each other, since each model is based on a separate and nonreducible (scientific) level n of the description that is related to the mentioned disciplines. The next row indicates the carrier and possibilities of decay or malfunction. The last row represents the number; L = relevant level at the four level architecture of intelligent systems, given by Radermacher (1996), Fig. 1, p. 4-7. The levels are: (1) = signal level, (2) = feature level, (3) symbol level, (4) theory level.

Entity and Level	Transforming Process Interpretation *as* ...	Relevant model for the transformation n → n+1	(Scientific) Level n of Description	Discipline	Carrier and Decay	L
Constellations, Processes, Situations			n=1 State Space description	Physics, Chemistry	Physical systems, decay by energetic processes	
	Processes to Signals	**Model 1:** Variables, Section of the World, Ordering Experiences		Measurement Theory	Measurement Devices, Sensory Physiology	
Signals			n=2 Time series, Dynamics, Stochastics	Comm. Technology,. Channels	Technical Systems, Distortions, Noise	1
	Signals to Symbols	Model 2: Pattern Recognition		Time Series Analysis (Cybernetics)	Technical and Neuronal Systems	1
Symbols, Signs			n=3 Reference, Symbolic Calculus	Semiotic, Syntax, Semantics	Surfaces, Decay by Dereferring	2
	Symbols/Signs to Data	Model 3: Aggregation of Signs, Application of a Mask		Model-building	Storage in Technology and Biology	2

Table 7: continued

Data			n=4 Computer Science Algorithms	Computer Science Hardware	Physical Carrier, Biological Carrier	3
		Model 4: Data Processing and Data Interpretation under a given Context		Approaches of Artificial Intelligence, Cognitive Science	Neuronal Net, self adapting machines	3
Information			n= 5 General Behavior of Systems	Biology Psychology Neural Science	Biological Systems, Nonclassical Systems with a semantic closure	4
		Model 5: Integration of Information into an already existing Knowledge		Theory of nonclassical systems, Theory of pragmatic information	Brains, nonclassical technical systems with a selfreferential semantic closure	
Knowledge			n = 6 Philosophy	Epistemology, Cognitive Science, Psychology	Individual Subjects, Society ?	

2.8
Consciousness

The last mentioned ability of handling knowledge internally can be subject of self-observation - *cogito ergo sum*. This self-experience tells us something about what knowledge we have in mind actually. Obviously we can distinguish it from knowledge that is potentially available but not actual „in mind".[23] Despite of knowing that it is dangerous to give a definition for consciousness, it may be attempted to do so: Conscious is shown by the ability to communicate about an adequate understanding of an outer physical world up to an adequate understanding of knowledge, within an internal and/or external framework.

It seems to be obvious that consciousness presupposes all the processes on the levels of physical constellations (physiology), of signal processing (sensors), of

---

[23] An example for this ability is the mental recording of music.

sign and pattern recognition (first calculation of outer world, H. v. Foerster), of data processing (for robots only![24]), of information understanding and of knowledge handling. Cognitive processes can be called all processes that presupposes the processes on all this levels, i.e. knowledge handling is to perform cognitive processes. If the cognitive processes became subject of themselves this may represent a precondition for consciousness, but whether this condition is a sufficient one, some doubts may be allowed. The concept of self-reference, expressing that an expression is referring on itself (it contains itself as an argument of a logical or syntactical operator), is only a formal representation of this circumstance, it does not found any sufficient condition for consciousness.[25]

3   Concluding Remarks

For any trial of Technological Enlightenment the above scheme distinguishing data, information and knowledge shows that there is no technological analogy for process of understanding, i.e. the transformation from information into knowledge. Therefore any speaking about knowledge technologies is a misleading metaphor, too. On the other hand it is true that the technological control of data handling and information transfer by sophisticated ICTs has empowered the societal and economic possibilities. It is very seducing to speak about a flip from quantity into quality, i.e. that the enlarged availability of information about everything at every time and at every place will create a new quality of Knowledge Society. The truth is that this information ability has substituted cash flows by data flows and in the realm of economic system this may be regarded as an information flow. This information flow can be re-transformed into cash flow - but this is done in only very few cases. Thus a theory of money becomes a theory of information flow.[26]

The necessary condition to acquire knowledge from information is given - of course - the context sensitive availability of useful information. But this condition is even a necessary, not a sufficient one. Thus the necessary filter functions, the ability to select from information flow must be developed further. For this filter function no other technological support those that is merely operating on data level (by indicator systems) is available until now. Here human wisdom, based on knowledge and experience seems to be the only remedy. Thus we cannot avoid the necessity of a human rational choice.

References

Bacon F (1597) Meditationes Sacrae. In: v Spedding J (ed.) The Works of Francis Bacon, Vol. 14, New York 1864

---

[24] I think that data processing is a typical level between the other levels, particularly for machines; in biological systems this level coincide with sign processing and pattern building.

[25] Cf. also Sullins III (1997), discussing the problem of Gödels Theorems and the possibility to build artificial life. The structure of the argument is the same.

[26] As conceived by Simmel (1900/1989) and C. F. von Weizsäcker (1971), p. 356-360, (1977), p. 265-269.

Carnap R (1974, 4. ed.) Der logische Aufbau der Welt. Scheinprobleme in der Philosophie. Das Fremdpsychische und der Realismusstreit. Berlin 1928; engl.: The Logical Structure of the World. Pseudoproblems in Philosophy. Berkeley, London 1967

Carrier M, Machamer PK (1997, eds.) Mindscapes: Philosophy, Science, and the Mind. Universitätsverlag Konstanz, University of Pittsburg Press

Chomsky N (1966) Cartesian Linguistics. A Chapter in the History of Rationalist Thought. New York, London

Churchland PS (1986) Neurophilosophy - Toward a Unified Science of the Mind/Brain. MIT Press, Cambridge Mass

Gernert D (1985) Measurement of Pragmatic Information. Cognitive Systems 1, p 169-176

Gernert D (1996) Pragmatic Information as a Unifying Concept. In: Kornwachs K, Jacoby K (1996), p 147-163

Haken H (1988) Information and Self-Organization. A Macroscopic Approach to Complex Systems. Springer, Berlin, Heidelberg

Heidegger M (1972) Time and Being, Basic Writings. Harper Torchbooks 1972. Dt.: Zeit und Sein. In: Heidegger M (1988) Zur Sache des Denkens. Niemeier, Tübingen, p 1-26

Kornwachs K, Jacoby K (1996, eds.) Information - New Questions to a Multidisciplinary Concept. Akademie, Berlin

Kornwachs, K (1988) Cognition and Complementarity. In: Carvallo M (ed.) Nature, Cognition and Systems I. Kluwer Akad. Publ., Amsterdam, p 95-127

Kornwachs K (1996) Pragmatic Information and System Surface. In: Kornwachs K, Jacoby K (1996), p 163-185

Kornwachs K (1967) Pragmatic Information and the Emergence of Meaning. In: Van de Vijver et al. (eds.): Evolutionary Systems. Kluver, Dordrecht p 181-196

Luhmann N (1991, 4. edit.): Soziale Systeme. Suhrkamp, Frankfurt a. M., engl.: Social Systems. Stanford Univ. Press, 1995

Pattee HH (1977) Dynamic and Linguistic Modes of Complex Systems. In: Int. Journal for General Systems 3, p 259-266

Radermacher FJ (1996) Cognition in Systems. In: Cybernetics and Systems: An Int. Journal 27, p 1-41

Ropohl G (1991) Technologische Aufklärung. Suhrkamp, Frankfurt a. M.

Searle J (1987) Minds and Brains without Programs. In: Blakemore C, Greenfield S (eds.) Mindwaves: Thougts on Intelligence, Identity and Consiousness. Blackwell, Oxford

Shannon C, Weaver W (1949/1969) The Mathematical Theory of Communication. Urbana, Chicago, London. First published in Bell System Techn. Journal 27 (1948), p 379-423

Simmel G (1989) Philosophie des Geldes (1900). Suhrkamp, Frankfurt a. M. (Gesamtausgabe Vol. VI)

Stehr N (1994) Arbeit, Eigentum und Wissen – Zur Theorie von Wissenschaften. Suhrkamp, Frankfurt

Stehr N (1994) Knowledge Societies. Sage, London

Sullins III JP (1997) Gödels Incompleteness Theorems and Artificial Life. In: Society for Philosophy and Technology - a quaterly electronic journal, Vol 2, Nr 3-4, p 141 (see also http://vega.lib.vt.edu/ejournals/SPT/spt.html

Turing AM (1950) Computing Machinery and Intelligence. In: Mind 59, p 236 ff

Weizsäcker CF von (1971) Einheit der Natur. Hanser, München

Weizsäcker EU von (1974) Erstmaligkeit und Bestätigung als Komponenten der Pragmatischen Information. In: Weizsäcker EU von (ed.) Offene Systeme I. Klett, Stuttgart, p 82-113

Weizsäcker C F von (1977) Garten des Menschlichen. Hanser, München

Zerdick A et al (1999) Die Internet-Ökonomie - Strategien für die digitale Wirtschaft. Springer, Berlin

IT-Security.
The Forgotten Dimension of Modernization

Otto Ulrich

„Today's systems of information technology offer neither adequate security against computer criminals, nor is the question of the vulnerability of our industrial society towards computer failure or misuse sufficiently clarified"
(Dr. Dirk Henze, the German president of the Federal Office of Security of Information Technology in Bonn)

1   Introduction

We cannot yet foresee today whether the security problem will in the end cause the failure of the information technology project. But it will be this area that determines whether the foundations of the information technology are firm and secure. Problems of information-technological security tend to be noticed only once damage has already been done. By means of some examples we wish to show the effects and the causes of some such occurrences.

- Due to a computer error a certain bank found itself in payment difficulties. The bank had to take up a special credit of 20 billion US dollars, for which it had to pay an interest of 10 million dollars.
- A telephone exchange in a large a German city broke down twice, presumably due to a software error: 10 000 customers were cut off on both occasions for several hours.
- The last few years have seen a increase in computer virus attacks, particularly on personal computer programs. The danger here is their uncontrolled expansion and the consequences of a possibly present damage function. A certain firm had to close down for six weeks and put in re-load all the data because a virus, due to the clumsiness of a operator, has rendered all the security copies unusable. The loss came to one million US dollars.

- A bomb explosion at outer glazed wall of a machine factory computer center caused a direct damage of 2 million DM. The damages caused by the consequent break-down in production came to 9 million DM.
- A flooding of a computer center caused by leaky roof during bad weather necessitated new wiring which took nearly three days. The damages came to 800, 000 DM.
- A frustrated operator caused several times a day inexplicable system crashes, each of which required extensive start-up procedures. The damages came to 1.4 million DM.
- It was programming mistakes which prevented the meteorologists from predicting the big storm in October 1987 in Great Britain.

2   Security of Data or of Information Technology?

On the way towards information society we must assume that computer criminality (misuse of personal data, financial manipulation, etc.), sabotage attempts, research- and industrial espionage using IT-systems and networks, attacks of hackers (break-ins into systems through international networks and their manipulation), computer viruses and other software manipulations are merely the top of an iceberg of largely unknown risks and security loopholes. They cannot be prevented and processed from the perspective of dealing with uncertainties. These unintended effects of IT have been considered so far insufficiently, one-sidedly, or ignored altogether.

The subject of *information technology* has appeared already 20 years ago. Originally directed largely at military needs, information technology has developed today into a broad inter-disciplinary subject, encompassing not only aspects of *data-protection* and *criteria of data-processing*, but also of *communication law*, attention to *security quality*, *norms* and *standards*, improved development facilities and tools, especially for *software production*, as well as investigation of methods of misuse and the corresponding counter-measures. One also talks of IT security, computer security, data securing, data protection-- these all being concepts which, though not synonymous, belong to the same context, namely that of safe information processing.

While *information security* is the broader concept, 'computer safety' and 'IT security' relate more to the technical aspects of security in IT systems and IT products.

Data-security is an older concept for information security; but its primary focus is the security of data.

Data-securing designates the sub-area of a safe storing of data.

Data-protection means the legal, technical and administrative rules regarding the processing and storing of personal data according to the data-protection laws.

However, to do justice to the complex phenomena of the modern dangers and the already recognized safety deficiencies of the multi-media applications, we need a broadened understanding of security, in the sense of *many-sided security*.

The cause of disturbances and break-downs of IT systems can be divided into three categories.

1. System-immanent errors
- Hardware errors.
- Software errors (also in the sense of deficient user-friendliness, or deficient robustness against user-mistakes).
- Fundamental deficiency of IT safety in public networks, e.g. the Internet.
2. Errors in the Set-up and Deployment of IT systems
- Errors in IT system protection (e.g. deficient access control and protection against a third party, as well as deficient securing of important data).
- Errors in the use of IT systems.
- Inadequate awareness of the problem.
3. Conscious Manipulation of IT Systems and Invasion from the Outside
- Conscious manipulation of data and IT systems.
- Invasion of IT systems from the outside for break-in, espionage and manipulation purposes.

Although many protective facilities and security measures are either on offer or already available in companies, a surprisingly small use is being made of them. This has the following consequences:

- Mistakes and negligence of employees, software-caused technical defects and software anomalies, together with hardware-caused technical defects remain the major danger zones.
- Mistakes and negligence of employees, software-caused technical defects and software anomalies will in the future assume an even greater significance for information security. Hardware-caused technical defects will, by contrast, decline.
- Only a third of all the firms views internal information security as a benefit. While companies do not consider information security to be unimportant, in many of them it does not have a priority.
- The greatest obstacle to improved information safety is deficiency in security consciousness, lack of competent employees and lack of money.

In summary we should point out that two factors are decisive:

- Faulty software
- The human error-factor

In software anomalies the IT security problem lies originally with the software manufacturer. 'Use-errors' and 'negligent use of IT security' are, by contrast, more home-made problems of the computer user. Though here it must be pointed out that the problem of use-error cannot be ascribed exclusively to the user, since software is often not user-friendly and error-robust.

3   Error-friendliness as Security Strategy?

It's true that the human factor is generally the weakest link in the application of IT, since it is the cause of most of the errors in the servicing and use of the IT systems. Here, however, we must also remember that software is also produced by human beings and is for that reason generally faulty. The current security strategy seeks to get around the immanent error tendency of the IT systems by security-directed 'hardening' of software and hardware, in such a way that the 'imposed' control technologies (security-technology controls error-prone basic technology) such as 'Firewall-computer', virus programs, user hierarchies, coding programs, etc., determine the user's right of access to the computer.

This basic idea determines all IT security measures with the consequence that the desired 'open, multi-media service society of the future', can acquire the requisite reliability and confidentiality only by imposing an additional IT security infrastructure upon the built-in national as well as global information infrastructure, which in turn immediately creates the question regarding the reliability and confidentiality of the IT security infrastructure.

An alternative IT security-strategy--one which does not exclude man but makes an active use of the peculiarly human capacity to learn from mistakes--proceeds from the following assumption: human beings always make mistakes. Because they make mistakes and thereby cause damage and accidents--mistakes may even play an evolutionary role--one cannot seek to eradicate mistakes, be it by means of error-proof technology and organisation or by means of legal and moral restraints upon the individual. Rather one must create a positive attitude towards mistakes. That, however, presupposes that one only permits error-friendly technology, that is a technology whose possible error-consequences are tolerable, so that people and society can learn from it. One may not apply any technology in which no errors are permissible because their consequences might be catastrophic. Establishing one's stance on the assumption that human beings make mistakes is an expression of an essentially new orientation. The criterion for technological production now become human, rather than technological possibilities.

4 Software Provisional Agreements as the Basis of 'Information Society'

The so-called Software provisional agreements are being currently implemented in economy, science, administration and everyday life as the basis of a digitally networked service society. This unreliability is one of the first experiences made by every user of the software systems. One would therefore expect that as they grow older systems become more mature and reliable. This, however, does not occur in the complex systems. Rather, the numerous mutual dependencies of the various components lead to counter-productive effects in error elimination. In the professional literature one assumes 1.5% to 2.5% of faulty instructions. And yet the proportion of specification- and design-error is in each case 30%, of coding-error around 40%. As regards development errors, a third of those are not yet discovered after the completion of the implementation phase. In the big software systems one must therefore expect a corresponding number of latent errors.

5 IT Safety Criteria

As a defence against these dangers, criteria for evaluating and constructing IT systems have been developed, which led to articulate the basic IT security goals. These generally recognised IT security goals provide today action-orientation for politics, economy and science. The following four IT security goals serve as the basis for all IT security conceptions:

- **Confidentiality**: designates security against unwanted inspection of information.
- **Integrity**: designates the character of a system which allows only permitted changes of the information which it contains.
- **Accessibility**: designates avoidance of unwanted retention of information or system resources.
- **Authenticity**: designates correspondence between a claimed identity and the real one.

As supplementary criteria of a *"many-sided or complete IT safety"* there are further:

- *Unobservability*: A 'communicative action', such as a telephone call, must be capable of being carried out without the knowledge of an outsider (the network manager counts as an outsider).
- *Anonymity*: A user must have the ability to get information and advice without revealing his identity, analogously to an anonymous purchase of a newspaper in a shop or an anonymous call to a SOS number.
- *Non-networking*: Several 'communicative actions', such as two calls between close relatives, may not be brought in connection with one another, since information gained in this way could destroy anonymity and unobservability.
- *Pseudonymity*: Anyone offering (information) services for pay to anonymous users must be capable of securing his payment in a safe way.
- *Indisputability*: Especially with respect to commercial uses there is a need for indisputable guarantees of genuine and punctual reception of a given information, such as an order or cancellation. Consequently, signatures of persons or of registered letters must be digitally or electronically reproduced.

6 Missing Long-term Accessibility of Stored Data

Registry offices, social and medical insurances, as well as statistical and financial offices, have, for some time now, tended to store their various data according to the new possibilities of information technology onto to the electromagnetic storage media (such as magnetic tapes, laser discs), according to the legal regulations, for 'eternity' – such at least is the intention.

It is known that the data stored today onto magnetic or opto-electronic bearers have a life expectancy of only some twenty years, where this must be interpreted in terms of 'half-value time': within that time at least one half of the bearers is no longer useful. (Further examples: the half-value time of paper less that 150 years old amounts to 100 years; of fax paper between two and ten years). It is further known that already today it is very difficult if not impossible, to reproduce data stored only ten years ago either because the components of the since obsolete technology have not preserved and /or because the know-how about the software of that time has been lost.

When, therefore, there is a need of permanent conservation of data one must take into consideration the temporary conservation capacity of a given storage medium.

7   Data-Mining

While data protection is still grounded on the data technology of the '70's which was based on large data-collections built according to homogenous criteria, today's information technology provides the possibility of searching heterogeneous data-collections for interesting data and relations. To combine these data which according to the current technical knowledge cannot be processed together, one uses statistical and heuristic methods, so as to link data-collections which stand in a logical relation to each other, and so gain information from them.

The areas suitable for the application of data-mining tools today are mainly large concerns such as insurances which, due to their distribution or production structures have large data-collections at their disposal and wish to gain from them further information about their clients. The same, however, applies to the analysis of data regarding currency developments in banks and similar data-collections.

The goals falling under the concept of data-mining combine bringing together to one person various data-collections – such as gathering all data in a whole concern to one person – with ordering of additional information to person-related data-collections out of available data-bases – such as extraction of group-specific behaviour data out of unspecified client data and ordering them on a probability basis to 'fitting' persons. The same method allows also to search in large production-data bases for errors and production inadequacies. The Internet, with its plurality of heterogeneous data bases provides rich material for many sides of data-mining.

8   The Economic Importance of IT Security

The structural significance of IT security for economy grows proportionately with the increasing spread and economic intensity of application of informational and communication technologies. The therewith connected increasing dependence of error-free running of data-processing means also the growth of dangers connected with the breakdown of or an attack on an information processing system. Data monitor calculates the damage caused world-wide by deficient IT security in a year at 16 million dollars. The importance of IT security as a risk minimising technical, social and organisational packet of measures has been dealt with in the introductory chapter and need not be further discussed here. The focus of this chapter will be rather the treatment of IT security with a view to digital markets.

The economic importance of IT security reveals itself in this connection in that businesses can only exploit the competition advantages of electronic commerce when they reduce the information technological risks. In this respect IT security appears as a confidence building factor without which both businesses and customers would shrink from taking the step into 'digital commerce'.

IT security as a confidence building factor reveals its economic significance only indirectly, since the market participant can simply avoid the 'digital economy' and carry out his transactions through non-digital markets. It is only by considering the cost-reducing and the therewith connected competition-effective mechanisms of the 'digital economy' that we can make clear the economic consequences of refusal to participate in economic commerce on grounds of IT security risks. This is why the importance of IT security in this context is being introduced through consideration of the economy of the 'digital market'.

The growing need of information technological security generates a demand for IT products and services, which initiates a market for IT security. This market and the competitive position assumed in it by European concerns will be considered in the second part of this chapter.

The emergence of digital markets reveals also the problem of suitable digital means of payment, which combine the principles of simplicity, anonymity and especially security. Just as real money requires wide-ranging security measures for all levels of its life-cycle, so too security must the systematically established at all levels of the 'digital money-cycle', if 'digital money' is not to be condemned to failure from the very beginning. Here protection against falsification is of particular importance.

Due to the world-wide spread of Internet, Electronic Commerce intensifies and further globalizes competition. The global inter-networking of computers which makes possible world-wide comparison of suppliers and prices, small costs of entering the markets and starting up business, as well as a potentially world-wide reaching of customers – all these new conditions of a global and digital economy generate significant pressure to accommodate not just for concerns but for entire national economies.

The importance of electronic commerce is strengthened by the explicit political will to promote electronic commerce by creating 'digital economy' adapted conditions of international commerce. With their "Framework for Global Electronic Commerce", President Clinton and Vice-president Gore set a clear signal for a consistent liberalization of electronic commerce. They are seeking to create through the Internet a global free-market zone without taxes and customs, a global virtual market, which follows only its own rules without state interference. This American-set tone has not remained without effect. In the developed industries we can observe clear liberalization tendencies due to the economic exploitation of the Internet.

Though Electronic Commerce finds itself at present still in an open development phase in which it still remains to be seen what economic activities can be meaningfully undertaken through Cyberspace, there is a great danger for the non-on-line concerns of losing those market shares which are used for information and transaction advantages of Electronic Commerce.

9   Economic Factor IT Security

The greatest entry barrier into Electronic Commerce is for both businesses and customers the deficiency of security in electronic network. Lack of Confidentiality, manipulate IT systems and legal insecurity generate risks which prevent confidence of businesses and customers in digital commerce. So long as it cannot be safely excluded that no unauthorized third party can read sensitive data, that authenticity of documents and the identity of the sender cannot be manipulated, are the potential risk costs of an invasion into a business transaction in data network too high. The economic disadvantages connected with the risks are viewed as higher than the advantages of Electronic Commerce for both the concerns and the customers.

The economic consequences of not using economic commerce on grounds of inadequate security is that the cost advantages of electronic commerce are not realized. The long term consequences would be competitive disadvantage of a large part of a national economy in global competition.

References

BSI 94 (1994, ed.) Computersimulation. (K)ein Spiegel der Wirklichkeit. Bonn
BSI 95 (1995, ed.) Patienten und ihre computergerechten Gesundheitsdaten, Bonn
BSI 97a (1997, ed.) Kulturelle Beherrschbarkeit digitaler Signaturen, Bonn
BSI 97b (1997, ed.) Mit Sicherheit in die Informationsgesellschaft. Tagungsband 5. Deutscher IT-Sicherheitskongress des BSI
BSI 98 (1998, ed.) Virtuelles Geld - eine globale Falle? Bonn
Bullinger H-J (1995, ed.): Dienstleistung der Zukunft, Wiesbaden
Deutscher Bundestag (1998, ed.) Sicherheit und Schutz in Netzen, Bericht der Enquete-Kommision: Zukunft der Medien in Wirtschaft und Gesellschaft, Bonn
Kersten H (1995) Sicherheit in der Informationstechnik. Einführung in Probleme, Konzepte und Lösungen, München
Müller G, Pfitzmann A (1997, ed.) Mehrseitige Sicherheit in der Kommunikationstechnik. Verfahren, Komponenten, Integration
Roßnagel A et al (1989) Die Verletzlichkeit der Informationsgesellschaft, Opladen
Tauss J, Kollbeck J, Mönikes J, (1996, ed.) Deutschlands Weg in die Informationsgesellschaft, Baden-Baden
Ulrich O (1994) Verletzlichkeit - Störphänomen der Modernisierung. In: Fricke W (ed.) Jahrbuch Arbeit und Technik 1994, Bonn

Threats of the Global Net-Internet.
The State of Development and Inquiry over the
Information Society in Poland

Andrzej Kiepas

One of the most important changes which we have to deal with nowadays is connected with increasing role of information and technology of it's storing , transforming and disseminating. The perspective of information society is one of many which one being discussed nowadays to the one about: post-industrial, post-modern, post-cultural society or the society of risk. Computing and development of multimedia technology are one of basics observed today in the world proceeding globalization of dependence. An indication of shown here processes connected with informatization is development of global internet. It exists from 70's but it's such big development is being the reason of technological nature can be observed in the last few years. Today internet is a supernet comprising tens thousands of data bases and nets also including such enormous as American On Line. Dissemination and development of internet was possible through elaboration the simple methods of accessing and using this net- world wide web. In Poland internet shown up in the 90's and during the last few years it's development is very easy to notice The observed growth of computers and internet organizations in the last years was 200% a year and according to estimation it is used by about half a million of people. It's here the citizens and young people in the age of 19-25. years who are majority. The considerable percentage are people with the high education and internet is still in Poland an exclusive expedient for today. However dynamic increase shows that in a short time we can expect many important changes in this area.

The interest about e-mail and World-Wide-Web is increasing. Popular magazines about computer problems and informational society (e.g. "PC World Computer", „Enter" and others.)are being published.

In 1995 Polish Association of Internet Users was created.This association disseminates knowledge and information about internet. But in the sphere of inquiries was not many initiatives concerning information society. It's possible to give attention to some of theme, for example:

- books published under the edition of L.W.Zacher where presented are world's problems of information society and new conceptions, for example: A. Toffler's, D. Bell's, J. Nuisbitt's and others;
- "Transformations" magazine also includes many articles refering to these problems but it especially refers to situation on the world;

- in 1997 Center of Information Society was created at Silesian University in Katowice. It is controlled by L.W.Zacher. Unfortunately because of financial regards it's activity is restricted ;in 1998 was organised a conference referring to problems of information society;
- Ministry of Communication organised in 1996 a conference about perspectives of information society development and Polish Academy of Science organised in 1998 a conference about problems of information and internet development;
- science of culture's representatives -K. Krzysztofek (Institute of Culture - Warsaw) and T. Miczka (Institute of Culture-Silesian University in Katowice)
- representatives of economic science -problems of innovation- E.Okoń-Horodyńska (Economic Academy in Katowice) and representatives of Institute of Administration of Technology (Jagielonian University in Cracow).

But what's most important there's a lack of institutions which in a continuous way is absorbed by problems of information society. The inquiries have in connection with it scattered character and interests about it are concentrated in the high developed countries. There's also a lack of settlements whether Poland will be creating it's own strategy of development information society what was one of the reasons that relatively well developed sector of electronic industry was not able to defy competition with the west.

Discussion about problems of information society shows different and important subjects in this matter but also attitude and position marking changes and development trends that can be observed. The role of extending and developing infosphere is differently marked and in general it's possible to favour four characteristically statements:

- optimistic-showing the positive consequences perspective resulting from dissemination and marking of information society in the global scale; is emphasized here is the increase of possibility of communication in the world scale, democracy in the local, international and global scale;
- pessimistic- they stress most of all different dangers that we have is deal with today; emphasize developing technology as unprofitable occurrence for culture and man; advancing instrumentalization of cultural world and life of a man is treated as important threat touching the border of human identity and not only the quality of his life;
- neutral-they treat all techniques ,including techniques of storing ,transforming and disseminating information as in itself axiological neutral. In itself they are neither good nor bad because it's intentions of it's user what matter and decide about it;
- statement acknowledging ambivalent character of some techniques: it acknowledges that technology in itself is ambivalent and only partly depends from our way of using it and partly from itself; technology is not a neutral expedient but independently from good or bad intentions of it's users there are always positive and negative consequences (Krzysztofek 1997, Kiepas 1987).

The most important meaning seems to have the last two statements because in comparison with the first two they have deeper character at least as it says about the ways of concerning the same technology. These statements are also characteristic for traditional and modern philosophy of technology. The normative turn which we had to deal with on the turn of 60's and 70's marked an important change in the way of conceiving technology, a change bending a step in the side of accepting the fourth from shown statements which is today the most typical one for the modern philosophy of technology (Ropohl 1991). However the third statement is still very popular especially among the representatives of different detailed science in the engineering technological environments but also in the instance of common opinions concerning technology as itself in general. However it is not realistic in a large degree what can be easily noticed looking closer at proceeded technology's development in human's life. Development of different technological expedients is not only different possibilities creation for proper and dependent from human decisions and activities but it also states human's life environment clearly determinates what kind of choices and activities are involved.

We have similar situation in this instance with internet. It creates positive possibilities being the tool of global communication but also dangers and negative consequences. We're going to concentrate on the last one with the consciousness that they are not the only one and that's why this occurrence which is internet shouldn't be too much magnified and demonised.

The negative consequences and occurred dangers can be divided in two basic kinds:

1. The one which are a consequence of particular way of using internet -here are positive consequences (e.g. growth or fastness of sent information ,possibility of accessing it, etc.)but also negative - the last one will be a consequence of different kind of abuses that we have to deal with in instance of global internet;
2. The one that are a consequence of proper international using of internet as a data collection and international but in consequence those activities have negative effect resulting from functioning and creating proper cultural patterns and communication culture; to this group of consequences -also positive and negative in general-belongs most of all social and cultural character consequences connected with creating global informed culture.

Considerable part of consequences belonging to the second group have non-international character because they are the result of incidental consequences of particular international activities or simply are the result of it's mutual teaming up so they occurred as a result of proper processes.

To the first group of negative consequences belongs also the one that refer to using global internet to dissemination pornography. It is not a problem that is restricted only to dimension of individual behaviour. It happens more often that people use internet in work and they spend their time looking through the pages „for adults". There are cases of dismissing people out of their works for using internet that way. On the basis of inquiry that was made in USA among bosses of 200 establishments about 20% out of them have punished for incorrect using of internet. About 1/3 of firms follow their employees and are interested in what are they visiting and what kind of materials are they taking from internet. That kind of vigilance is particularly conducted in bigger firms and generally the different firms interest about the way of using internet by their employees increases. They are also interested in sent and received e-mail (Martin and Rzeźnicki 1998). Those occurrences growth and make different kind of problems of technological, economical, legal and moral nature.

Thanks to internet visited are not only pages for adults but also shopping and private cases are being made. It is also a problem and that's why at least by toss-if it is not made systematically-takes place monitoring of how employees are using internet during their work. To this group of consequences of abusing the net can also be included occurring more often instances of dependence from internet what mainly refers to young people who are using internet in excessing way that it leads to their isolation from the world. The indication of threats that appears nowadays is occurrence of infoholism and abusing different kind of computer's technologies and excess of information. They can't usually answer the question "what for they need so much information? "and can't also use those information properly. For those people computer is becoming their world replacing them relations and references to the real world. Running away from the real world they usually direct themselves with the will of pleasure. They also usually have narrowed intellectual horizons and restricted fields of interests. It is also connected to the quality of information sent in the net. Democratisation of access causes that a considerable part of information is devoided an important value but also it's estimation isn't easy. It needs special knowledge ,sensitivity and in consequence criterions allowing to estimate what is important from what isn't. The net itself doesn't supply that kind of criterions and further more it doesn't create proper axiological sensitivity. It needs to be possessed from other source and because of this invasion of information young people who are in natural way devoided axiological basis are threated especially.

Here also appears a generation problem which has widen cultural dimension. The culture of modern societies becomes to be most of all a culture of presence because patterns of behaviour and activities are obtained not from parents and elder generations as also not from existing social, legal and moral order but often from mas-media including information from internet. This change of form of intergenerational transmission states to appoint M. Mead as a postfigurative culture with elements we have to do with nowadays (Mead 1978). Examples of abusing the net also refer to politics. Many big and small terrorist groups already have access to internet so it can also become a tool of international terrorism. It can also refer to possibility of propagation different ideologies -not always with humanistic premise but also nationalistic, fundamentalistic ideologies.

A separate and enormous problem is the matter of evince and a secret and possibility of obtain the access to different data by average user of the net. In spite of existing different protections they are almost always possible to break.

The second group of threats concern social and cultural dimension. It is not completely independent from the first one anyway-it is the opposite because they are mutually connected. The area of transformation and threats connected with development of information society is quite different and wealthy and that's why we restrict here only to some problems referring to cultural communication ,policy in global and local scale and ethics. At the basis of every culture and proper way of it's communication in the past always lay certain pre-theoretical opinion on the world. Modern information culture seems not to be connected to necessity of having that kind of basis. It is a culture of momentary states which sense disappear with the stimulus that has caused it. The following stimulus doesn't need to have something to do with the earlier one ,doesn't have to appeal to the same ,,code" or the same opinion on the world-it can be completely temporary and can build it's own senses in a different way. The world of momentary structures -proclaimed by postmodern prophets- is a world in which tendency to world's subjectivization initiated in modern tradition is brought to the extremes. Evincing since modern times reversal of earlier relations between subjective and objective mind in consequence leads to the whole world subjectivization, it's sensualization and instrumentalization. It also has connection with difference between cultures of high and low influence context. As L. F. Groff emphasizes in his regard: "In the sphere of interhuman communication the main thing is difference between cultures of high and low influence context. In the cultures of high influence what refers to many traditional cultures- East, Arabian, Latin, Russian and others- in communication the most important is situational context and mutual interhuman relationships. Until there won't be established relations between people leaning on a trust ,it isn't possible to sit down and do business. In the opposite to the low influence cultures -like in USA -where people are directed for a purpose ,want to get to a business right away and achieve understanding not often giving proper amount of time for socialization and getting to know the second side" (Groff 1997). It is not only a problem referring to intercultural relations but also a problem of global Americanisation of world culture under the influence of internet (it is only one out of many expedients). L. Groff claims -to be sure-that this globalization doesn't have to lead to homogenization of modern culture (Groff 1997). Despite the meaning that is attributed to cultural variety it's maintence doesn't seem to be completely obvious. Occurrence of globalization carries with itself danger of cultural colonialism and doesn't have to lead to maintenance and increase of role of national and local cultures at all. It is now difficult to state univocally what tendencies will turn out to be dominant in the future. Optimists think that there is no threat of occurrence of technostructure but it isn't certainly sure. Development of information society -Often in spite the declaration-doesn't tolerate existing partitions. If necessary it replaces it with new once, e.g. on societies rich in information and poor in this regard. Trial of searching own way of development by the ones who stayed in the back may also be unsuccessful because modern high technology is difficult to master and produce in preinformation societies. Vision of whole democratization of interhuman relations in the global scale might turn out to be a utopia-similar to vision of computer or interactive democracy. It is also possible to use particular techniques but also it is possible to make different manipulations by those who are disposes of those expedients. Choices of politicians are in considerable degree dependant from the image of particular candidates created by mass-media. Does the elector have possibility of checking genuine of those images? (Krzysztofek 1997) Practically none. Representative democracy will have sense ,,when society will have possibility to choose the programmatic elites so actually multimedia empires. The meaning of internet and techniques of sending and storing information also evinces in the sphere of education. Processes of moving away from the Gutenberg's galaxy towards picture of culture radically changes dominating here earlier rationality giving information - which is often also an entertainment - don't give the key to understanding the world but only proper to receive such or others impulses. Cultural environment of modern man becomes to be technology but it is produced and supervised by different experts and that's why increase of interactivity and creative possibilities from the modern receivers of mass-culture postulated by optimists may turn out to be a myth. This activity may only be apparent and in fact reduced to sensualistic response to different impulses.

In the world of internet we have to do with elements of virtual reality where noticed might be characteristic for modern culture displacement of the process of communication referring to:

a)   transition from „'bookish' world of acquaintance to it's navigating" -in reality it means lack of cultural context and enter to the world of temporary structures
b)   replacing transmission of information with interactivity
c)   modification of the preparation into inclusion (Miczka 1997)

Virtual world -just like the whole world of internet -may entice giving opportunity of subordination it to the will of individual and on the account of lack of financial conveyor which was earlier an important limitation but it can fulfil with the expense of mutual isolation of the individuals and the lost of cultural and social identity. Enter into the world of internet and virtual reality may be attractive but there is also a threat of losing freedom and identity.

In today's world of threats and risk increases the meaning of responsibility which meaning is not confining only to the sphere of ethics.

Responsibility in today's information culture becomes dependent from knowledge, information and technology and not only from sensitive of active subject;

- must be based on a trust not only to the sent and received information but also to the expedient of it's displacing and transformation; trust to a computer as an information conveyor and expedient of it's transformation is the base of decision succoured by a computer;
- decentralization of the ties and globalization of dependence characteristic for information society indicates in some way that those ties adorn the net's character; independence of net's type the consequences of activities are a simple result of a cause and effect dependence; there's also many non-international consequences;

Traditional way of understanding responsibility was connecting responsibility with causation what nowadays isn't that much univocally sure because of degree of connection of different individual activities the way that their consequences exceed beyond that what was in the possibilities field of activity of particular individuals and what allow isn't contained to the end in fulfilled by them social role. In the situation of net's dependence responsibility becomes to be co-responsible but it's undertaking requires not only conscious subject but also defined degree of world's clarity where they will be moving in. Plunged in the world of temporary structures the internauts may have some problems in this matter and that's why the question whether they can be the subjects of co-responsibility acquires today so dramatic dimension. It is also said today about the need of breaking modern tradition in which world and different areas of man's practical activity where divided according to the rules of instrumental mind but there's also lack of ethical dimension. They are not immoral in themselves but they are concentrated on what is useful and that's why they loose this ethical dimension from the field of view. It's recovery is today connected with responsibility and co-responsibilities popularization and it is a challenge for modern internauts - but they must realize that information obtained through the net are not insufficient to take this responsibility in a conscious and effective way (Kiepas 1997).

Appealing to responsibility is the final appeal to humanity which is threatened today by the development of technological civilization. Information society will bring new and often by appearance of attractive forms of threats. Only responsible man will be able to manage the challenges which he has to manage because in other way it won't be his problem anymore.

References

Groff L (1997) Communication between cultures: Assistance in international understanding and cooperation in the world of today. In: Zacher L W (ed) The problem of the information society. Warsaw, p 26-27 (only in polish available: Komunikacja międzykulturowa; pomoc w międzynarodowym zrozumieniu i współpracy we współczesnym świecie. In: Zacher L W (ed) Problemy społeczeństwa informacyjnego)

Kiepas A (1987) Introduction in the technical philosophy. Katowice (only in polish available: Wprowadzenie do filozofii techniki)

Kiepas A (1997) Ethic in the information society – New problems and challenges. In: Zacher L W (ed) The revolution of the information society. Warsaw, p 64-167 (only in polish available: Etyka w zinformatyzowanym społeczeństwie - nowe problemy i wyzwania. In: Zacher L W (ed) Rewolucja informacyjna i społeczeństwo)

Krzysztofek K (1997) The information society under revolution of the telecommunications. In: Zacher L W (ed) The revolution of the information society. Warsaw, p 53-62 (only in polish available: społeczeństwo informacyjne i rewolucja teleinformatyczna. In: Zacher L W (ed) Rewolucja informacyjna i społeczeństwo)

Martin J, Rzeźnicki D (1998) Observation oneself. Will be oneself in danger? In: PC World Computer. January, p 110-112 (only in polish available: Obserwuję cię. Czy jesteś w niebezpieczeństwie?)

Miczka T (1997) Functional Opportunities. From the point of view of the communication. In: Zacher L W (ed) The problem of the information society. Warsaw, p 36-45 (only in polish available: Rzeczywistość wirtualna- w perspektywie komunikacyjnej. In: Zacher L W (ed) Problemy społeczeństwa informacyjnego)

Ropohl G (1991) Technologische Aufklärung, Frankfurt a.M.

Scientific and Technological Progress, Democracy, Participation and Technology Assessment in Russia

Vitaly G. Gorokhov

1 A New Understanding of the Technological Progress and the Concept of the Sustainable Development

The present engineering activity and designing in the second half of the 20th century are characterized by substantial changes, which allow to indicate a new non-traditional form in their development. This of course does not mean that their traditional forms are disappearing and go off to a background. They continue to function effectively and obtain, however, a new quality by engaging in a wider context. For example, the creation of a car today does not represent simply a technical development of the car but includes also a creation of an efficient system of servicing, development of a network of motor roads, production of spares and so on. The building of electric power stations, chemical plants etc. requires not simply an accounting of the "external" ecological situation but the formulation of ecological requirements as a source data for designing. All this ruts forward the new requirements both upon an engineer and designer and the representative of technical sciences. The influence of the human being, society and nature is so great that their social responsibility before the society is immeasurably rising. Therefore, it causes a problem of reorientation and of changing of an engineer and designer thinking.

Technical activity, which is characteristic of humanity since its earliest development, turned into engineering when it began to align with science (the regular application of scientific knowledge to technical practice) or, at least, began to operate with scientific vision of the world and when engineers acquired a professional status and a special engineering education appeared. Since the Renaissance, modern culture has essentially been based on design and oriented at creation, invention, and technical progress. Ancient cultures were canonical, being based on time-honoured traditions. For this reason, inventors in a modern sense could not have existed in those times, although there were inventions, and there was also a design function latent in canon which allowed the canonical culture to evolve into the modern design culture. It is this latent design thinking which is actually "philosophy" in these technologies of the canonical cultures.

There were two main ways of the technological reflection in the ancient cultures, which have an implication in the modern culture:

- idea of the maintaining of existing social and natural order and practical technology in the ancient China (Needham 1977; 1984), striving for harmony between society and nature in ancient India;
- the way of aggression, the war against (or attacking) nature to capture the nature, the formation of the myth about machine - a social mega-machine (Mumford 1967) - aggressive seizure (capture) of the human (social and natural) environment.

The second aggressive approach is a characteristic feature of the modern technological development in the first half of the 20th century. "Unfortunately, man differs from animals not only in that he can think but also in that he can perceive, create and reproduce incorrect information, believe in myths and illusions and, which is the main thing, act according to incorrect information. Our technogenic civilization is one of the illusions as we believe that can everything" (Gorshkov, Kondratyev, Danilov-Danilian, Losev, pp21, 22).

In the modern technology there are the several different "philosophies" of technology:

- systems philosophy and project management: a methodology of systems engineering (design here, there and everywhere - a principle of the design culture);
- social technology and organizational design: design of the society (local and global social structures) - necessity to aware of the boundaries of the design culture in the social engineering and social design (society is not an artificial system, but natural-artificial system);
- biotechnology and necessity of a new engineering ethics (understanding the internal boundaries of the technological development conditioned by the biological nature of human beings);
- information technology and AI-investigations - a new possibilities of the conceptual design ("expert systems") - strengthening of the theoretical dimension of technology: design as a research and research as a design (erosion of boundaries between research and design);
- environmental technology and a new "philosophy" of the sustainable development - (understanding the external boundaries of the technological development for the mankind in the biosphere).

Modern Systems Design	Traditional Engineering Design

1. The object of design

system of activity = large-scale man-machine system	engineering (technical) object is means of activity: artifact, machine, device etc.
environment as an "external" element of designed system	-
individual ("unique") object of design	standard engineering object - multiplying
the activity in making large-scale system becomes to an object of designing ("design of design")	-

2. The process of design

implementation activity is not designed, but organized; it is impossible to take into account all the parameters and particularities of functioning of a large-scale system	manufacturing: according to a given project in existing production one can make an artifact corresponding to the project
evolutionary design: the design of system does not stop after putting it into operation; one correct different stages of the realization of project on the base of study of regularities of functioning of designed subsystems	single engineering cycle has to be stopped after putting of the engineering object of this type into operation (one begins design of engineering object of new type)
development and gradual improvement of the existing system according to the project	design "all over again"
design without prototypes: oriented on realization of ideals, forming themselves in methodological sphere	design by prototypes, which already exist in the engineering practice or in the nature
direct relation with consumer ("participation design"): in the case the design itself becomes to a source of formulation of project themes; the systems criteria formulates designer himself	task of design is formulated by customer (mediator between manufacturer and consumer); he determines also the criteria of the quality of the finished product

Modern Systems Design	Traditional Engineering Design

3. The sphere of application

all fields of social practice: production, service, consumption, management, education and etc.	only industry

4. The means of design

using of knowledge, methods and representations of the whole complex of modern sciences (social, natural, engineering sciences, mathematics)	using of methods, which are elaborate mainly in appropriate engineering science
dialogical character: a comparative programs, projects, plans; oriented not on a unique theory, but on the methodology, which ensures the unity and variety of consideration at the same time	monological (monotheoretical) primarily orientation on the basic engineering science
systems method, representations and notions (system orientation).	natural-scientific knowledge and representations (orientation first of all on physical picture of the world)

The engineer must listen not only for the voice of scientists and technical specialists and for the voice of his own conscience but for the public opinion as well, especially when the results of his work may influence the health and way of life of the people, affect the monuments of culture, disturb the natural environment etc. When the influence of engineering activity becomes global, its decisions cease to be a narrow-professional matter, very often they become a topic of general discussion and sometimes - a topic of conviction as well. Although the scientific and technical development is reserved for specialists, the decision to be taken on such a type of designs is a prerogative of the society. No references to the economic and technical and even the state expediency can justify the social, moral, psychological, economic and other damage impaired by some designs. An open discussion of the designs, explanation of their advantages and disadvantages, structural and objective criticism in the public press, social examination, promotion of the alternative designs and schemes are becoming an important attribute of the present life, an inevitable condition and consequence of its democratization.

The present engineer, designer and researcher in the field of technical sciences should joint the world and national culture besides all other via a deep study of the history of their profession. As far back as the start of the century, P.K. Engelmeyer wrote: "Very often and fairly the engineers complain that other spheres do not want to recognize their important significance, which must rightfully belong to an engineer ... But are the engineers themselves ready for such a work? ... because of a shortage of the general intellectual development, the engineers themselves know nothing and do not wish to know about the cultural importance of their profession and consider the discussions of these matters a useless waste of time ... From here a task rises before the engineers themselves: to increase the intellectual development inside their own medium and basing on the historical and sociological data understand the entire importance of their profession in the present state" (Engelmeyer 1913, p112).

2 Politics, Progress, and Technical Professionals in Russia and Soviet Union

In the Russia before October Revolution were many different professional societies independent from state as the real guarantors of the democracy also in the sphere of the technology assessment. We can give only one example.

The Society for Promoting Advances in Experimental Sciences and Their Practical Application at the Imperial Moscow University and the Imperial Moscow Technical School named after Professor Chrisian Ledentzov deserve special mentioning. It was functioning in Moscow between 1909 and 1918. Prof. Ledenzov of the Imperial Moscow Technical School bequeathed 100 000 roubles to set up this society on the indispensable conditions that "all members of the Society, irrespective of their sex, social status, academic degree and nationality should promote the Society's aims stated in the charter. This is mainly done through grants for the development of the discoveries and inventions that can, with the minimal capital investments, bring greater advantages to the majority of people. These grants should be used to help implement the discoveries and inventions but not trail behind them as prizes, subsidies, medals, etc." Ledentzov stipulated in his will that "the Society should cooperate with Messrs. Innovators not so much by granting money as by promoting the maximally profitable utilization of their discoveries and inventions. The conditions of such a cooperation should be agreed on in written form beforehand. An any case, part of the profit should go to the Society's special fund intended exclusively for the practical implementation of discoveries and inventions" (Annals 1910, p. 10).

The Society's charter stated:

1. The Society's aim is: (a) to promote discoveries and research in the natural sciences; (b) to promote technological innovations and improvements; (c) to test and implement scientific and technological discoveries and inventions.
2. This aim can be reached by (a) giving advice and instructions and discussing the submitted projects, issuing expert opinions about scientific and technological discoveries and inventions, distributing grants for the implementation of scientific research and innovations, setting up laboratories and other similar organizations; (b) publishing the Society's works, establishing libraries, carrying out public readings and discussions, organizing museums and exhibitions; (c) working to allow people recommended by the Society to

participate in special works in the teaching institutions and departments of the Imperial Moscow University and the Imperial Moscow Technical School; (d) making the best possible use of discoveries and inventions on the conditions agreed beforehand with innovators so that part of the profits go to the special fund for the promotion and implementation of discoveries and inventions: some other part of the profits should be use to strengthen the Society's finances. The correlation between the two parts should be decided on by the Society's general meeting; (e) awarding medals, prizes and diplomas for scientific and technological research and discoveries" (The Moscow, p. 91). The Society channeled 10 000 roubles out of its extraordinary balance sheet to set up a library and to replenish it annually. It was regarded as a form of promotion of research and discoveries and a base for expert and consultative efforts.

Today, we can learn a lot from the Society's experience. What is very important is its main idea to give money not to ready-made products but to ideas that may bring practical advantages in the future and, in this way, effectively to support gifted researchers and inventors. To be efficient, a similar society could be founded as an international organization. This alone could ensure a genuinely *independent* assessment of technological projects. We should be concerned not so much with isolated technological assessment as more care about the social-humanitarian and ecological sides of any project. An assessment of these aspects should take into account specific social and cultural features of all countries.

Unfortunately, this society was existing only before 1918. All-Union Association of Engineers survived until 1929, when the political situation change dramatically. Very frequently, errors and breakdowns in technical operations were taken to be intentional acts, motivated by ideological opposition.

These development destroyed previous efforts to build a common professional self-identity for all engineers. With the change in political focus, the publications in the magazines and journals of the profession carried militant editorials that praised the "great engineer of human souls" - "great engineer of the social reconstruction", "social architect of genius" - namely, comrade Josef Stalin. Articles with titles like, "The Face and Pole of Anti-Soviet Engineers", attempted to whip up fears about technical professionals secretly organizing to undermine the revolution. In 1933-1934, N. Bukharin's contributions to this campaign published in *Social Reconstruction and Science*, a magazine dealing with the problems of scientists and engineers. In one of the article Bukharin stressed pathetically: "The army of militant builders is on the march with flags flying, the sounds of the International reverberating in the air, and the leader of the army is revolutionary Field-Marshal, is teacher in the gray greatcoat, setting the pace with his will of iron - JOSEF STALIN" (Bukharin 1934, p. 21). Today we aware of where Stalin was leading the country. But at the time many people took these developments as an inspiring prospect for the large-scale remolding society.

A representative example of this new mode of operation was a project to link the White and Baltic Seas - the Byelomoro-Baltiisky Canal. It was a gigantic social experiment carried out on the bones of more than a hundred thousand people. Among its goals was that of reforming the lives and consciences of all who worked on it.

Engineers experienced Stalin's care and concern in other ways as well. The first special designers' offices (Special Constructor's Bureau = SCB), called "sharashki" in low colloquial, appeared during the 1930s and employed the forced labor of convicts. As the design offices became widespread throughout the country, they were described enthusiastically in political speeches and the press as "beginning". As a result of Stalin's "concern", thousands upon thousands of lives were lost. But many people decided to focus only upon the new, limitless prospects for socialism and the rapid place of material achievement. "Reforming" was meant to help engineers "overcome the conditioned reflexes of the capitalist era". The process of drawing sabotaging engineers to socialism went on "under high social pressure accelerating the thinking and nervous reactions" and reforming engineers "biologically".

Reporting on workers at the Byelomoro-Baltiisky Canal, one official commented, "When at liberty, they used to slow down the pace of their work deliberately. There they could be over-subtle, throw dust in one's eyes, quote someone as an authority, while here they are convicts, and there is no throwing dust in security men's eyes. The Party and government decision about the terms of construction is irrevocable, and all kinds of false rumors are out of place". True, the conditions were hardly fit for work; laborers were placed in confinement, "at the world's end" with the shortage of metal and cement, and without a single skilled worker and with no "complete technical design's documentation". The job did not leave any time for thinking. "The plan, the inexorable working plan, turns gradually into the supreme law that must equally be obeyed by security men, engineers, thieves, bandits, and prostitutes" (Rykachev, 1934).

Such was the "paradise at the world's end, under the skies of Karelia and GPU (State Political Department) guardianship" where the engineers experienced "a great joy of life". Those were the engineers, the "general technical headquarters of Russian capitalism", presumed saboteurs, who found themselves in the ranks of the enemies of the Soviet government. They were arrested by the GPU, pleaded guilty, and expressed readiness "to begin working for Soviet power", i.e., to build the canal (Rykachev, 1934).

There was a Russian way of the industrialization.

"Russia's way differs greatly from those of other states with centralized systems. Capital to build modern industry was obtained like under a usual marked-based system - by robbing the people. Producers were alienated from the means of production proclaimed to be the "property of the whole people". In fact they became the property of a capitalized state where all decisions were made by a handful of "leaders".

Industrialization was created out with the use of most advanced technologies. Soviet specialists were trained in Germany and the USA. Major enterprises were built mostly in old industrial regions where nature was deformed greatly. Developed were usually Russia's former outskirts - the emergent socialist republics.

Economic growth (to catch up and pass the West!) became the main task. Communism was the ultimate goal but a closer and more realistic aim prompted by the idea of world revolution and hostile surrounding was to create a powerful military machine. During the Second World War industries were moved to old industrial areas east of the Ural, in Western Siberia and to Central Asia and Kasachstan.

After the war a new stage of technological modernization began when whole factories were brought from defeated Germany. Huge investments were made in the development of new kinds of weaponry and the arms race began when military technologies were rapidly developed and an enormous military-industrial complex was established. All that was financed by keeping the people on the subsistence level.

Such situation could not last long in the conditions of peaceful coexistence to which all nuclear powers were doomed. Peaceful coexistence presupposed competition, and Nikita Khrushev and the Soviet leaders who followed him to raise the living standards and at the same time money to maintain the military machine. Natural resources were barbarically exploited for that purpose.

The campaign for developing virgin and fallow lands was followed by the efforts of cotton-growing base and then by the large-scale oil and gas production in Western Siberia. All those efforts, however, did not prove effective enough in competition with market based systems, and centralized systems were ousted and on their ruins states with transition economies appeared" (Gorshkov, Kondratyev, Danilov-Danilian, Losev, p. 18).

The broader lessons from this chapter in the history of the Soviet Union can be summarized. It is impossible to anticipate everything "from the top". Some reasonable combination of planning and a self-regulating market is needed. Centralized manipulation of huge resources, transferring them by orders and decrees (without outside social appraisal of experts and regardless of public opinion) is not advantageous, but detrimental to nature, society, and human beings. This is demonstrated by the consequences of attempts to change the flow of rivers and crash programs for building nuclear power station (NPP). There is no way to make people happy from above against their will and without their agreement and active participation.

3 Ecology and Technology Assessment in Russia today: Environmental Monitoring as a Means of the Participation

For the modern time development of the Russia appropriates the follow description: On the whole, centralized systems are being ousted peacefully, and there is hope that the process will continue to be peaceful. Market systems are just means of adapting basic principles of life to the existing civilization. No mathematical models or computer systems can establish prices and offset balances better than the free market. Abandonment of the free market means loss of accuracy and greater unproductive expenses. The biosphere can be said to be built on principles similar to those of the free market.

Russia came to the transition period with a structurally deformed industry whose absolute enterprises produce low-quality goods. The transition to be free prices, the cuts in subsides to unprofitable enterprises and also the reduction of the armed forces and arms production have resulted in an economic recession. Money is needed for a new sharp rise in production on the basis of advanced technologies, and Russia obtains the money like other countries do: by lowering the living standards through higher prices, trade and fiscal machinations. The method works till the bulk of the population reaches the subsistence level. Then political difficulties usually occur and the domestic market shrinks to the minimum. In such conditions the government should do two things: stop the drop in the living standards and do it quick (any delay spells a turn back to centralization). That is precisely what is called "shock therapy". The sooner the ends the sooner investments in production begin and the living standards begin to rise. Under the pressure by the military-industrial lobby and other groups the former Supreme Soviet is doing its utmost to impede it. The reforms should be continued energetically and the state subsides to unprofitable or unnecessary enterprises should be stopped as they remind one of the labor of Sisyphus. Any delay is politically dangerous and increases inflation" (Gorshkov, Kondratyev, Danilov-Danilian, Losev, pp19 - 20). This text is written 5 years ago, but it is actually also today.

One of the interesting example of the technological development give a sphere of the nuclear energetic.

Before Chernobyl catastrophe was concerned that the nuclear power stations have a good operational reliability and are a very good energetic perspective for the next century. They have no more ecological consequences as another, for example, steam power plants. It was a hopefulness that a problem of a mediaeval alchemy of the transmutation of the natural elements of the world is solved and we receive soon a paradise in the Earth. But after Chernobyl catastrophe is this dream disappeared. And today more and speculated about closing the nuclear power stations in the different countries especially in Germany. Besides the problem of the recycling and the end disposal of atomic wastes is not solved (neither technically nor scientifically) and the costs of these recycling and disposal was not calculated in the costs of the building of the nuclear power station. Neither forecasts of the (not only ecological and for human health but also financial) consequences of these possible accidents in the nuclear power stations. So it is that the development of the nuclear energetic is the most striking instance of the unpredictability and uncontrolled consequences of the scientific and technological progress which arose from "aggressive" line of the technological development.

Application of atomic energy for peaceful purposes was a by-product of the military-industrial complex in defense of the development of the atomic weapon (Menshikov 1998).

Application for peaceful purposes of the for military goals developed technique to bear the impress of the secondary by-product: as an military action has an peaceful application a definite percent of the victims of war (of an accident, of pestilence, of circumstances).

Our civilization would be inconceivable without the many things brought about by engineering activity. Engineers and designers have brought to life what once seemed incredible and fantastic, but they have also developed sophisticated means of mass destruction. Although technology is ethically neutral, the engineer cannot be indifferent to its application. Leonardo da Vinci was so worried that his inventions might be put to undesirable use that when describing his idea of an underwater apparatus, he did not describe his method of remaining under water and possible time of staying without eating. This he did not publish or reveal on account of the evil nature of men, who would practice assassination on the bottom of the seas by breaking the hulls of boats and wrecking them with all on board. This example of the moral stand was bequeathed by Leonardo da Vinci to future generations of engineers (Arsakanian 1984, p. 132). No assertion of national, economic, or technical expediency or higher scientific interests can justify the moral and material damage that can be inflicted on people and environment. Scientist and engineer is responsible to the living and future generations for his activity and its results.

After Chernobyl catastrophe the scientific view of the world is changed. It is understood that:

- is need to have an independent experts for technology assessment,
- men's knowledge and scientific prediction are limited,
- for creators (not living near from power station), staff of the nuclear power station (who understand that this is a dangerous object) and population (living near from this station but not understand what dangerously it is) situation is quite different,
- is need to inform the population and the political set about the normal or extraordinarily situations in the Nuclear power station and near from it. This is very important to organize independent (from the operational organizations, designers and emergency organizations etc. but qualified) environmental monitoring of the radiation situation near from ecological dangerous objects. For democratization of the technology assessment of the nuclear safety the international information change and the organization of the free visibility of date play important role. On the basis of the full information is it possible the realization of the free choice for individuals as a main principle of the democracy. This means a possibility of the participation of the citizens in the decisions-making by technology assessment not only after creation of the technique but also before and in the process of the design. Today also is discussed a problem of the *"Machbarkeit der Technik"* - not all as possible to make necessarily must be realized (Lenk 1994).

The results of the engineering activity surround us everywhere, and the engineering approach has made inroads into scientific, social, and even humanitarian spheres. Social engineering, biotechnology, engineering economics, and so on are coming into being. They influence medical practice via medical instrumentation and pharmaceuticals production. Engineering activity affects the environment both on a regional and on a global levels. The effect of the scientific and technological progress on society and nature became global in the end of the 20th century, producing a host of acute ecological problems. As engineer is not merely a specialist in some area of technology, he deals with both nature, which is foundation of life, and people. Modern engineering therefore poses problems of social responsibility, intellectual honesty, and professional ethics.

References

Annals of the Ledentzov Society for Promotimg Advances in Experimental Sciences and Their Practical Application (1910). Moscow (Russian)

Arsakanian ZG (1984) The Problems of the continued connection of Renaissance and the "Age of the Scientific Revolution". In: The Problems of the History of the Natural Science and Technology, N 2 (Russian)

Bukharin NI (1934) Science and People: Heroic Symphony. In: SORENA, no. 2 (Russian)

Engelmeyer PK (1913) Thanks of the philosophy of technology. Bulletins of polytechical society, no. 2 (Russian)

Gorokhov V (1990) Engineering: Art and Science. Moscow: MIR Publishers

Gorokhov VG (1997) Peter Klimentjewich Engelmeyer. Mechanical Engineer and Philosopher of Technology. 1855-1941. Moscow: Nauka (Russian)

Gorshkov VG, Kondratyev KYa, Danilov-Danilian VI, Losev KS (1994) Environment: From New Technologies To New Thinking, Moscow: Federal Ecological Foundation of the Russian Federation

Lenk H (1994) Macht und Machbarkeit der Technik. Stuttgart: Reclam

Menshikov V (1998) Russia with the atomic energetic and without it. In: Russia in the Environment: 1998 (Analytical Yearbook). Moscow, International Independet University for Ecology and Politology (Russian)

Mumford L (1967) The Myth of the Machine. Vol. 1. Technics and Human Development. New York; Harcourt Brace Jovanovich.

Neeedham J (1977) Wissenschaftlicher Universalismus. Über Bedeutung und Besinderheit der chinesichen Wissenschaft. Frankfurt a. M.: Suhrkamp

Neeedham J (1984) Wissenschaft und Zivilisation in China. Frankfurt a. M.: Suhrkamp.

Rykachev Y Engineers by Byelomor Project. Ogonyok Library, 1934, no. 16\787, p. 6, 3 and 13 (Russian)

The Moscow State Historical Archive, recourd Group 224, inv. 1, File 1 (Russian)

Transformation of Society and Values in Slovakia

Pavel Fobel

1 Introduction

The social change, which occurred in the countries of Eastern and Central Europe at the end of 2nd millennium, penetrated expressively into all the substructures of the social system. It started a movement within not only the political and economic mechanism but also existentially influenced the socio-cultural and spiritual sphere of life. Although that break generated a new form of mental behaviour and expectation connected with the longing for prosperity, freedom and consideration all over the world, it simultaneously also evoked well-founded anxiety and uncertainty concerning the complicated constitution of the new social system. As well as each principal change, it brought up more essential questions with regard to alternatives of global and inner-state effort. It initiated ambitions of integrated processes within Europe; the most convenient political forms of both economic and military direction and progression; national aims of emancipation; ways of democratic arrangement. In the former Czechoslovakia and then Slovakia, the crystallisation of political potency and tendencies (even without particular risks arising from economic development in the revolutionary year 1989) decided to use a historical chance and to form the separate state. So, responsibility for independent development was assumed, too. Relatively in a short time, in historical terms, the dynamics of these movements deserved admiration, mainly in order to its calm development, inspite of many risks and anxiety which resound still now. This evaluation initiated the necessity for expert discussions and theoretical analyses - sociological, politological, but also interdisciplinary discussions on a European nation, as well as the ecological, spiritual, and historical aspects of the Slovak nation.

Particular matters of the economic, political and spiritual conditions in Slovakia have evoked many dilemmas of a systemic nature. As the existing researches and indicated analyses show, the transformation will be a longer-term process than it was supposed in the euphoric times of idealised visions. The evident disability of such a rapid change in the economic and social spheres - increasing prices, inflation, the lack of immediate unilateral solutions supporting either prosperity or social certainty – all these facts were reflected in some scepticism and uncertainty over the possibility of a positive change. Functioning reform became an existential constancy both of the system and the individual citizen. It has influenced the most sensitive spheres of life, its microworld of values.

The transformation and opening up of society in terms of freedom and democracy also brought new negative phenomena. These are allowed to originate within the undeveloped juridical and social framework and might be enforced like socio-economic and mental (nearly apparent and admissible norms of behaviour) forms: unfair self-enrichment, narcomania, corruption, malice, fraud, mafias, pornoindustry, assassinations and violence in general. Democracy is also threatened by the habits, stereotypes and conformity of the former regime, where autocratic and centrist-voluntaristic approaches dominated. They are reflected in the dichotomy of a powerful personality versus voluntaristic anarchy. The complexity of reform has even strengthened such a polarity because neither of its not fulfilling imagination of certain programmes for mental improving social situation nor of its offering an algorithm for clear society-expecting behaviour. Socialist industrialisation and urbanisation ("towns next to the factories") did not solve the problem of a civic society, too. Although the "from-above" planned voluntaristic, socialist industrialism and urbanism changed the social structure, they did not generate a principal reversal in the mental sphere to the virtues of industrial culture. Individual labour ethics and the principle of personal responsibility were absent. A citizen became "subject to" this historical process. Lots of open questions according to their number and urgency are gaining increasing tendency. However, these problems and difficulties of development do not eliminate the effort to enforce authentic democracy, economic prosperity; aims to ensure spiritual and social certainty as well as international consideration. A man's change, his thinking and anthropological transformation of values might be considered one of the most difficult tasks.

The economic reform brought the phenomenon of social differentiation, change of life style and social uncertainty. Evolution of unemployment has got an increasing tendency (during recent years it was 4% and in 2000 an average unemployment is expected to reach 20%). Simultaneously, the costs for active policy in the labour market are descending. From a total of 83 regions, in 10 the costs have already reached a limit over 30%. The structure of institutions, bodies and services for citizens was stabilised by the establishment of the state governmental and political structures originated in an independent Slovakia in 1993 as well as in its new territorial division in 1998. Small business unit and corporation stagnation, low standard of competitiveness in production, the high standard of insolvency and material costs cause the following stagnant effects, economic bankruption (or failure), exceeded employment and getting the sack, high-level unemployment concentrated in certain towns. New introduced programmes and technologies for production are economically demanding projects. The unsatisfactory amount of foreign investment, often determined politically, is able to solve pretentious social problems just to a certain extent. However, inspite of the high standard of professionalism and specialisation, the economy is not able, intellectually, to take control of the contemporary adequate demands on technique and specialised activities, to evaluate itself up to the possible intellectual potentiality, especially for developed specialised armament production. Although the post revolutionary humanisation caused a change within the programme, the moral credit of effects in the national economy increased, but it brought up no social certainty and a valuation which is full of contradictions.

If we evaluate citizens' opinions and statements towards democracy according to the economic transformation and comparison with balance against the past, we notice some antagonistic attitudes which basically reflect this process complexity, the psychic and social consequences of such changes. Research in IPSAS (1993) (in the first year after the birth of separated Slovakia) signalised uncertainty and disorder caused by the division of traditional values as well as deeply established roots of some attitudes which will influence the whole value orientation onto Slovak citizens for some time. At first, we have to take into account preferably such a reality that for 41 years – it means from 1948 to the revolutionary year 1989 – the average age of the 10 million inhabitants in former Czechoslovakia was under 41 years. It shows that about 65% of population ran through socialisation of the former regime. Some ways of consideration and behaviour, like leveling, etatism, comformism, egalitarism, planned industrialisation, political centrism with no individual responsibility and freedom, have taken its roots too strongly. They all penetrated both into public life and the private sphere displayed in the puzzled reception of the economic transformation in everyday life. In addition, long term habits and stereotypes were turned out into unproved steps within problem solution, so anxiety concerning future development was strengthened. It seems as if origin certainties granted by the former regime began missing. Only 6% of respondents think that our democracy has not suffered from insufficiency and 27% of them consider objective change was achieved quite well. However, 58% of respondents think our democracy is quite weak. It has not been decided so far if our system is more democratic or autocratic. According to sociological statements evaluating the state of our society, it is possible to claim that the scissors have been stretched out in greater distance between verbal respondents declarations supporting democratic principles and real citizens' behaviour.

As nearly each citizen of Central-Eastern Europe has lived in two regimes, he is at least able to differentiate qualities of both of them. The comparison of positives and negatives in the Communist regime shows that citizens' opinions are in better accord within the positives of the former regime and they are less similar in determination of the negatives. Most of the respondents (Czech Republic, Slovakia, Romania) evaluated the positive social certainties of the former regime (Mihálikova 1996). The negatives are seen mainly in the sphere of political discrimination (67%) but also in the following democratic values – inability to express their own attitude, the unlimited power of only one party, the standard of the ruling elite, restrictions on freedom and religious life. The peculiarities of Slovak democracy are also expressed in the next data: up to 93% of respondents think there are too many discussions and on the other side too few solutions to eternal problems. Most of them are also convinced about the strong concentration of political power. Such negative features of politicians like careerism, profiteering and corruption are strictly criticised.

The realised economic reform, with its processes and ways, interfered with man's value microworld. Nearly one third of respondents is not included in privatisation, and do not trust the private sector. Just 13% desire its predomination. Up to 50% of respondents want the private sector to play the same role as the public and co-operative one. According to the results within valuing privatisation as an act of historical fairness, the prevailing opinion is to give back the properties without harmed public interest and determination of their price.

Some other sociological researches and data of analyses as well as foreign experts valuations have indicated a culmination of social, but mainly recession and the retardation of economy with the ability to be stabilised and revived in two or three years. This prognostic data supposes the following actual risks and their influence in the social and moral sphere of a citizen's life. Despite that, the results of researches both abroad and at home confirm the secondary influence of economic factors onto the change of preferable political values in Central-Eastern Europe. So, it is very difficult to assert the necessity of "a powerful hand" regime. It would have to secure a transition towards the economic prosperity to deny democratic freedom. However, people learned patience (because they had nothing to spare in the past) and were disappointed. They want to be convinced about the positives of expecting freedom and certainty in the new society. In this process also civic helplessness plays its role according to the complexity of the solved questions: HOW TO SOLVE THE PROBLEMS? There is also a lack of positive experience and reputation by social world (Búterová, Gyárfášov, Kusk 1996). Passivity has got also a gnoseological background concerning weak information, unevaluated alternative thinking, value disorientation, lower ability to perceive circumstances, to think up consequences, etc. A lot of our and foreign researches, indicate the population of Central-Eastern Europe have not found an optimal model of participatory civic culture. They are convinced that in democracy, the dominant role is in the hands of experts, then professional politicians and at least citizens. However, only 2.8% of Slovak citizens think that contemporary politicians follow universal values. The act of apathy or the phenomenon of non-defending democracy can be perceived as a refusal of active participation in politics. In order to be specific, we are introducing three examples enabling the perception of the problems within several specific life spheres and active ways how to solve them in Slovakia.

2 Socialist Industrialisation and its Economic Consequences in Slovakia

Slovakia is being geographically and variably well kept country including its country side. So, environmental matters are perceived as being important. Socialist industrialisation, technological era brought into its valuation some new elements which have displayed during the period of democratisation and have shown negative influence. Entrancing commercialism in order to using natural sources, undeveloped tourism, the lack of means to revitalise areas in danger, changing the owners relations and only gradual perceiving responsibility to countryside, imperfect legislative – those are brief characteristics describing contemporary situation. Socialist industrialism accepted neither the difficult global technological, ecological parameters of Western Europe nor rational arguments concerning the possible consequences and risks of introducing projects. The construction of nuclear power stations, zones of danger, coefficients of emission because of out of date technology cause a problem with neighbouring countries, political dilemmas of an international nature and criticism according to the accepted standard in Europe. The thunder of political and civic discussions has become the form of economy versus ecology, respectively ecological arguments versus state interest. In spite of many unfavourable ecological factors, foreign criticism, low financial resources, the process of transformation has created some presumptions for active environmental politics. The systematic monitoring of the ecological situation, politicians will deliberate with foreign experts participating in the solution and the evaluation concerning the situation in Slovakia, global participation in realising international projects and grants, publicity and public information about the situation on the Internet, media, care of experts preparation of Universities, high schools, a functioning independent environmental agency across Slovakia, high standard of specialised culture, readiness and specialists looking toward a solution to the ecological problems – these are certain positives and more optimistic expectations. In terms of civic responsibility the projects of Healthy Town are also remarkable. They are introduced by city communal policy to take care of the quality of the life in towns. They have become the guarantee of the improving environment and at the same time the quality of life, as well as the values of "The House of Europe". There are some predominant questions concerning civic and political discussions and dilemmas – construction, shutting down nuclear power stations, licensees and the environmental sense of the Gabčíkovo dam, a candidature for winter Olympic Games, loans and introducing modern production technologies, their economic and social influence, reform of legislation with regard to European parameters, new proprietary relations, etc. Opened dialogues, expert preparation, international communication, rational and systemic publicity profile, intellectual reactions to problems and the behaviour of citizens, their mental being in general. For example, picture 1 shows the actual situation about the environmental level, the active effort to perceive our ecological situation, ways of giving information and ecology publicity on the Internet.

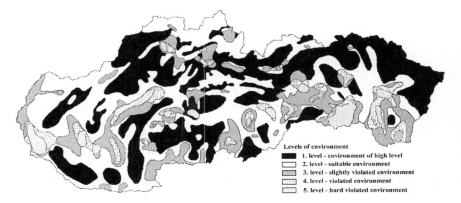

Picture 1: Levels of Environment in Slovakia - Levels of Danger

3   Development of Science and Technology in Slovakia

While in the surrounding countries (Czech Republic, Poland, Hungary) some principal, mainly legislative changes in the sphere of science were made in the 90´s, the development of science and technology in Slovakia was suppressed except between 1990-1992. No creative material or purpose was accepted, no legislative act which could correct any principal problem according to the development of science and technology. In such a situation there is no rational way out of the more than critical situation because of general lack of finance including the scientific and research sectors in Slovak Republic. The efficiency of the public, university and business sector can be brought up to a standard common in mature small countries within the European Union only through the following system changes: to find new ways to finance science and technology from the state budget; to accredit institutions of research as well as their experts who are supported from the state budget; to set up new legislative rules and to join research institutions including universities into the profited and non-profited groups. There is also necessity of competent authorities and institutions, a respective legislative framework- the goal of non-governed scientific and university subjects has been initiated for a longer period. A peculiar sphere is presented within the system of changes with regard to both the scientific and technical equipment at universities. In spite of yearly university evaluations, regular accreditation and about 70% of means from budget chapter, science and technology have been financed through purpose-made finance. Circumstances have become critical because of decreasing costs. However, the university law of the year 1990 defines universities as institutions of education, science and art we cannot say that all of them have to be top scientific institutions. As a result of that it is necessary to provide detailed diversification of the faculties to accredit them.

4   Information Technologies and Universities

Certain positive qualitative shifts with regard to information systems are noticable within the economic institutions in Slovakia. In each faculty and university independent information systems are realised. The following extended chances to work with the Internet (its use on an academic grounds is free of charge) are : to create separate pages for each university subject; to arrange separate library information systems, the possibility to use it for preparing time-tables, agendas, entrance tests, to systemise and present publication activities according to European norms, to manage and communicate within the workplace, etc. Introducing the latest information technologies into the academic world brought up some serious problems. The majority of Europe has joined the academic network TEN 155 (including the Czech Republic, Hungary, Poland). Slovakia has not been included so far (see picture 2). We are considering to form the Board of Cabinet for informatics supposing a deputy of the Slovak Department of Education. We feel lack of informants for applied information system at universities, because of the financial covering, as well as a consistent system for introducing programmes. The next problem is to keep experts at universities. On

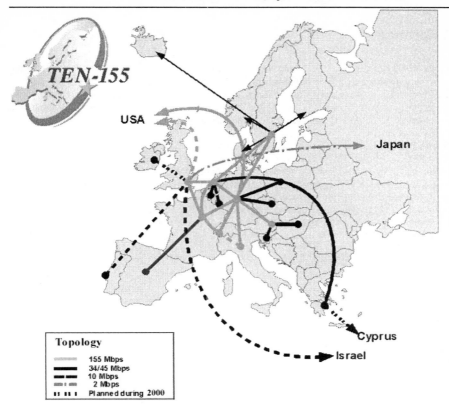

Picture 2: Pan European Internet TEN-155 which join educational, scientific an research institutions started to use from December 1998

the other side, we appreciate the initiative of MICROSOFT with its licencing policy for schools. As to the programme SELECT – it offers the possibility to buy SOFTWARE for 10% of the regular price. We hope immediate discussion on supplying the SELECT programme will lead to an agreement between the firm MICROSOFT and the Department of Education and will legalise SOFTWARE (it means the teacher may use it legally both at work and at home for his own purpose).

To assume – the development of information systems and technology within the academic world is perceived at universities as a strategic aim for full compatibility with the European Union and the demands of the information community. The fact that in 1997 private persons bought two-times more computers than all the schools in Slovakia is strange.

Our era is specific in a global and national context. Slovakia perceives this very intensively from the aspect of civic and social self-reflection. The XX. World Congress of Philosophy points out at the contradiction of our era which has developed the crisis of the human. "It is necessary to show and to discern alternative ways to which a human submits his behaviour and simultaneously he must take responsibility for them." (Fobelová, Fobel 1999)

References

IPSAS (1993) Institute of Philosophy of the Slovak Academy of Sciences (1993, ed.) Open Society. Bratislava (only in slovak available: Otvorená Spoločnosť)

Miháliková S (1996) Conception of Democracy and Democratization (Concerning Some Connections with Theory and Praxis). Sociology, vol. 28, no. 5, pp 415-430 (only in slovak available: Koncepcie demokracie a demokratizácie (k niektorým teoretickým a praktickým súvislostiam))

Búterová Z, Gyárfášov O, Kusk M (1996) Development in Slovakia after Elections in the Light of Public Opinion Polls. Sociology,vol. 28, no 5, pp 431-460 (only in slovak available: Povolebný vývoj Slovenska vo svetle prieskumov verejnej mienky).

Fobelová D, Fobel P (1999) Philosophical Mission and Appeals for Education on the Threshold of the Third Millenium (Reminiscences on the XXth World Congress of Philosophy in Boston). Organon F, Philosophical magazine, vol. VI, no. 1, pp 102-108 (only in slovak available: Poslanie a výzvy filozofie k výchove na prahu tretieho milénia (Reminiscencie na XX. svetový filozofický kongres v Bostone)).

IV The Judgement

Rationality and the Use of Language. Technology Assessment for Shaping the Knowledge Society

Armin Grunwald

1   Introduction

The task of technology assessment (TA) is, generally speaking, to investigate and to reflect consequences and impact of technology and technicalisation for the society and its individual members. Its development since its very beginning in the sixties has followed the path away from ex-post investigations of existing technology towards reflections *ex ante* in order to enable society to influence technology at early stages of the development. Such considerations *ex ante* are unrenouncable elements of shaping technology in society (Bijker and Law 1994; Rip et al. 1995; Grunwald 1999a). Applying TA to the challenge of shaping the knowledge society[1] demands for combining both the *ex ante* and the *ex post* approaches. The development towards an increasing use of digital technologies is, on the one hand, going on since the early seventies and has already affected many areas of society. This allows the social sciences to perform ex post analyses and empirical research on the impact of those developments. On the other hand, many chances of networking and of global communication are still at the stage to be *promised* to become realised in the medium or far future (Tauss et al. 1996). TA in this field, therefore, can be leaned on substantial experience but is still facing a large field open for shaping and reflection *ex ante*. In particular, there are time delays in introducing the information technologies in different zones of the world, for example between the United States, the Western Europe and the Eastern Europe countries. This situation demands for learning effects from one another. It challenges TA to perform international and intercultural comparable studies (see below, section 5).

Up to now, the concept of rationality has not widely been used in TA approaches (Grunwald 1999b). In this paper recent methodological work on TA, starting from the concept of pragmatic rationality (Grunwald 1999b, 2000a, 2000c) shall be used to deal with certain aspects of the challenge of shaping the knowledge society. In order to proceed in this way some basic issues of ‚rational TA‘ have to be presented briefly (section 2). In applying this approach to the

---

[1] The discussion around the concepts of information or knowledge society is not touched in this paper. Compare Mittelstraß 1998, Janich 1998a.

present discussion on the knowledge society emphasis is put on the investigation of the use of language (section 3). The role of rational TA in critically reconstructing the language involved is exemplified by (a) uncovering some presuppositions hidden in the public discussion concerning the idea of a technological determinism (section 4) and (b) by critically reviewing the frequent talking about internet-based democracy, touching the fields of regulation by the state and the chances and limitations of public participation (section 5).

2   Technology Assessment and Pragmatic Rationality

Rational TA claims to analyse and assess both descriptive-epistemological as well as normative and ethical aspects of new technologies and their impact on society using rationality criteria (Grunwald 1999a). The concept of rationality is understood in a procedural way: the rationality of propositions and the rationality of requests or recommendations is checked by uncovering the argumentation chain supporting that proposition or request and the investigation of the validity of that argumentation. Rational propositions or requests are those which can be shown to be valid „for everyone" – they transcend the mere individual beliefs and constitute trans-subjective validity. Actions and decisions are to be designated as rational if they are undertaken upon a basis which can be shown to be rational *in its normative and descriptive parts* (Rescher 1988, Gethmann 1996). The reflection on the consequences and the impact of the technological advance includes, therefore, the methodological reconstruction and critique of the results (theory of science) as well as the judgement of the generalisability of the normative criteria for the assessment (ethics). In this way, pragmatic rationality is not restricted to a positivistic understanding of rationality or to the „technically halfened reason" (Habermas) but also covers the ethical dimension.

The well-known expert dilemmata (for example, Nennen and Garbe 1996) are tackled by this TA approach *not* by renouncing the claim for argumentative rationality in favour of mere mediation and bargaining. The conviction is, instead, that there often is a *lack of rationality* leading to the expert dilemmata. Accordingly, the way of dealing with these dilemmata is to add as much rationality as possible to the conflict types leading to the various dilemmata. The problem that each expertise could be devaluated by a counter-expertise (expert dilemma of the first kind) shall be resolved – as far as possible – by reflection on the premises and presuppositions of the diverging expertises and by the assessment of their validity (Gutmann and Hanekamp 1999). In this way, it cannot be guaranteed that the conflict may be resolved completely; however, it will be reduced to its kernel and freed from avoidable and misleading controversies. The normative question for the legitimation of evaluative expert judgements[2] (expert dilemma of the second kind) shall be handled by recurring to models of resolving moral conflicts which have been developed by the professional ethics (Gethmann

---

[2]   In some former TA approaches this dilemma should be avoided by a „decisionistic" sharing of responsibility between sciences and politics. In the participative TA it is, on the contrary, claimed to resolve the problem by including citizens to the process of decision-making (Renn/Webler 1996).

and Sander 1999). The general conviction of rational TA is, in summary, that society should, as far as ever possible, rely on the chances for solving conflicts *by argumentation*.

Where are the sources for such a rationality assessment and upon what basis could it be leaned on? In the pragmatic concept of rationality the formation of reliable communication and action structures manifests itself amidst the multiplicity of participating players in a society (*Teilnehmerperspektive*, Habermas 1988; Hartmann and Janich 1996). Neither references to systems rationality (Luhmann 1990) nor an exclusive focus on isolated players as is the case in some economic rationality models can be justified *as a general and non-reductionistic model of rationality*. The pragmatic concept of rationality does not refer to some transcendental subject or other instance outside of society. The concept of and criteria for rationality are "constructively" created by and in society. Therefore, this understanding of rationality is a *culturalistic* and not a transcendental one (Grunwald 2000a). It does not lead to culture-invariant statements but uncovers the rationality implicity acknowledged and followed by society. This basis allows some kind of long-term reliability and stability because inherent societal rationality standards do not vary rapidly - in contrast to the short-ranged acceptance behaviour (Grunwald 2000a). Two main consequences shall be mentioned briefly:

(1) Even rational actions and decisions in the sense mentioned above cannot ensure their success: rationally acting is acting under the conditions of risks; no guarantee can be given even in the case of an action with maximum rationality included;

(2) The validity claimed "for everyone" does not imply a strong universalism. Who is meant by "everyone" must be cleared contextually: a small group, the population of a region or a state or the whole of humankind including future generations. It depends on the answer to the question who is affected by the decision under consideration.

The concept of rationality as proposed here is a normative one *though* self-constructed in society. It implies self-obligations and obligations for other people in order to act rationally in the pragmatic sense mentioned above. In this way, rationality is a *self-construct of society for the purpose to transcend individual beliefs towards transsubjectivity in order to enter the level of well-founded collective actions and decisions*. It is used to optimize the success of collectively relevant actions and decisions, to enable societal learning processes and to allow the balancing between continuity and flexibility in society.

3 Technology Assessment and the Use of Language

The first stage of a rationality assessment is to analyse the use of language in the field considered, because (1) the meaning of words, terms and notions often is unclear and leads to misunderstandings preventing transparent decision-making and (2) the validity of propositions and requests only can be assessed by using the means of language available. Forming a catchword and introducing it into the public discussion often guarantees the leadership in that field *though* the particular meaning of the catchword and the connotations involved might be unclear or even contradictory in itself. The particular use of language covers normative expectations, fears, aspirations etc. Talking about the knowledge society may serve as an excellent example for the necessity of uncovering the hidden or implicit meanings transported by the words used. For example, consider the following notions being frequently used in the public discussion:

- information flood,
- information garbage,
- digital revolution,
- data highway,
- Nintendo-generation.

It seems easy to imagine that these notions used in contexts of *describing* aspects of the knowledge society are carrying additional, evaluative and normative, semantics. Such terms, therefore, do not only describe something but include normative expectations about the future (positive or negative). Societal transformations of any kind (including the transformation to the knowledge society driven by advanced information technology) can be described by using very different terminologies:

- Far-ranging transformations may be denoted as revolutions with a positive connotation (as often done by talking about the „digital revolution"); one may appreciate the technological advance as modernisation and expect societal progress as a consequence of the technical innovations.
- They may be described, in contrast, as elements of a deep crisis bringing important achievements and standards of society into danger; it may be lamented about the „dark side of innovations" (Schumpeter), the price to be paid for the modernisation may be highlighted, a loss of traditions or values established in society so far may be feared.

Attempts to describe aspects of the knowledge society by means of language does not, as a rule, lead to value-free descriptions. Instead, normative expectations or fears are transported within such descriptions. The spectrum of talking about the knowledge society reaches from attempts to re-establish former types of an uncritical belief in the technological and scientific progress up to the pessimistic rhetorics of cultural decay. For example, assumptions concerning the impact of future information technology on democracy are fluctuating between aspirations of an interactive democracy allowing a maximum of participation (the ‚paradise' vision, compare section 5) and the expectation of the state evolving to the „big brother" being able to completely control its citizens (the ‚hell' vision). An example from the past is to consider the introduction of the private television channels in the eighties. The terminology used by the promoters was characterised by increasing the plurality of perspectives, deregulation, enforcing private initiatives, freedom etc. Today the same development is often characterised as cultural decay, as decreasing instead of an increasing plurality, as quality decay of the programmes to the lowest level available etc.

These examples show that choosing the terminology of description is an act carrying normative implications, hidden in metaphors, visions or cultural backgrounds: descriptions are more than mere descriptions. Innovations always cause winners and losers – normative evaluations of such transformations are hidden in terminologies describing them. Visions and metaphors have large influence on shaping technology and the public discussion (Mambrey/Tepper 2000), especially in societies heavily influenced by mass media. The task of rational TA is, therefore, to uncover such normative ideas and to analyse them with respect to their rationality (Grin/Grunwald 2000). This rationality reflection should cover, at least, the following fields:

(1) the *descriptive dimension*: what mechanisms of technology development are presupposed and how can such presuppositions be justified (compare section 4);
(2) the *moral dimension*: what normative ideas are included in the terminology choosen and in what way can they, if ever, be justified from the moral point of view;
(3) the *political dimension*: what ideas about legitimate political decisions are transported by the terminology used, how do these ideas meet the societal consensus etc. (compare section 5).

4 Technology Assessment and Technology Determinism

As a little case study illustrating the discussion above the underlying descriptive assumptions of talking about the knowledge society with respect to the model of technology development presupposed will be discussed and assessed in the following. Terminologies may be characterised – in the respect of human and cultural development through history – by using a more active or a more passive wording. Within a passive wording it is presumed that the knowledge society will come in the future as an inevitable course of the societal evolution. The only way of dealing with this, assumable pre-determined, evolution, thus, would be to give society the chance to be prepared for this self-running development. Behind this image there are ideas of a „technological determinism" (Ropohl 1982, p5ff.) assuming that technology development is following its own mechanisms and cannot be shaped or steered by society. Society then could only *adapt itself* to the ongoing technological progress, and, in the best case, could compensate some of the negative consequences by political regulation: adaptation instead of shaping.

Ideas of such technology determinism have also been followed partly in the early TA discussion. Since the eighties, however, approaches are in the foreground which emphasize the chances of *shaping* technology by and in society (for example, Bijker and Law 1994, Rip et al. 1995). In main parts, these approaches are based on ideas derived from the social constructivism (Bijker et al. 1987).

These approaches are talking about the future in a more active-sounding terminology.[3]

Returning to the ongoing discussion on the knowledge society it seems that in this field the technology determinism has not been overcome up to now. Speeches of politicians and industrials are often carrying expectations that the knowledge society will come independent of what is done at the present time. Sometimes they sound as a renaissance of technology determinism (sometimes assuming this development as to lead to the golden future or to destroy our culture and traditions, see section 2): „The importance of embedding technology in society as a process open for societal shaping ..... is under-estimated dramatically at the present time" (Mambrey 1998, translation A.G.).

Indeed, the determinism seems to be plausible to some extent and is supported by several special issues of that field. At first, the development of information technology is dominated by the rules of economy. This seems to disable authorities of the state to regulate or influence the further development of information technology and its adaptation by society. Secondly, this effect becomes more important in a globalised world (with the information technology itself being a driving force for globalisation). Who could (legitimately) steer a really global development? Thirdly, in the public opinion the development of information technology seems to be dominated by some companies or even singular persons. But if, as is assumed in this view, for example, Bill Gates would be dominating the development to the knowledge society the question arises, who could persuade, steer or force Bill Gates to do the best under aspects of a „common good"? As a conclusion, there are some serious arguments against an over-estimated optimism with respect to the malleability of the knowledge society.

If the way to the knowledge society would indeed be pre-defined and not open for societal shaping the question remains if the societal adaptation being left as the only way to deal with this challenge should itself be an active or a passive one. Sometimes a mere passive adaptation is favoured:[4] „Spätestens 1998 muß in Deutschland freie Fahrt für Informations- und Kommunikationtechnologien gelten. Bis dahin sind noch vorhandene administrative Hürden und Hemmschwellen abzubauen und Anwendungs- und Produktperspektiven zu ent-wickeln" (Rüttgers 1995). This directive by the former German minister for technology aims at a complete deregulation to eliminate any regulative restriction for information technologies. In this context it is very interesting that at the *first* place it is mentioned that administrative hurdles have to be removed and, only at the *second* place, it is postulated that new application perspectives have to be developed. Nothing could better uncover the hidden determinism – the „normal" way of proceeding would be to have application ideas first and then to reflect on possible requirements for removing existing restrictions.

---

[3] Sometimes it seems that these approaches may over-estimate the chances and possibilities for shaping and understate the limitations given by existing social structures, traditions, economic driving forces etc. But following this suspicion would lead far beyond the task of this paper.

[4] It seems to be reasonable to maintain the original German sound for this and some other quotations. For the reader who does not understand the German text its meaning should become clear from the enveloping discussion.

What's about the rationality of the assumption of a technological determinism? The way of analysing this question is to uncover the argumentation supporting this hypothesis. Various attempts from the social sciences, the systems theory or by using biological analogies are centred around the theory of evolution. Evolutionary approaches seem to be dominant at the moment. They try to explain technology development by using the evolutionary principles of variation and selection (Basalla 1988). If such interpretation patterns would be valid (the question for the criteria of validity itself seems to be very interesting in this context) any reflection aiming at shaping technology would lose its sense at all. Technology development would follow its own paths, uninfluenced by TA or technology policy. Society only could try to predict this pre-determined developments in order to get some information how to optimise the process of adaptation to this self-running evolution.

Many arguments have proven the assumption of a technology determinism to be false (compare, for empirical reasons, Bijker et al. 1987, Bijker and Law 1994, Rip et al. 1995; for theoretical reasons Ropohl 1982, Grunwald 1998). The main argument seems to be that *even if we accept the idea of technology determinism we have still to decide.* Technology is always developed in the perspective of participants in certain processes of research, design, planning, production, management etc. (*Teilnehmerperspektive*). At each stage of these plural processes decisions have to be made, according to the level of influence and responsibility of the persons involved. The idea of technology determinism *does not free* the participants from performing these decision-making processes. It is not helpful at all for decision-makers if someone talks about technology determinism and the impossibility to shape technology because the assumed determinism *does not imply in what way the decision-maker shall act.* He or she is completely left alone with the necessity of choosing between several options.[5] Instead, a theory is demanded to support decision-makers in doing their job. This might be a kind of rational decision theory applied to the field of technology development (for some ideas, compare Grunwald 1999b).

Instead of an „irrational" idea of technology determinism leading to the assumption that one cannot do anything but to adapt society to the „digital revolution" (see the quotation from Rüttgers above) it should be tried to establish a model of technology development starting at the perspective of participants. This attempt leads to the model of a *directed incrementalism* (Grunwald 2000a).[6] It maintains the rationality in terms of means and ends dominating the actor's behaviour at the micro level without assuming that the ends of acting are fixed. Permanent reflection on the goals and the means implemented to arrive at the goals leads to *incremental but intended* changes of direction in the development, of the goals as well as of the measures to reach the goals. This change, however, does not occur upon the basis of chance events and does not show an erratic

---

[5] Obviously, it is not required that every option which may be imagined is included in the set of options among the selection has to be made. Also choosing one option among a limited set of options is a decision and implies a certain aspect of shaping the future.

[6] This model may be very close to the idea of technology being a „seamless web", which has been brought into the discussion by the social constructivism (Schwarz and Thompson 1990, Schwarz 1992). In comparison, the perspective of setting goals and purposes and their reflection seems to be emphasised in my approach.

behaviour; instead, the development makes it possible to get closer to the envisaged aim and, simultaneously, to take into account the short-ranged flexibility requirements (which are leading to the incremental changes of direction). Directed incrementalism, therefore, fits very well to the concept of pragmatic rationality and seems to be very suitable to explain technology development as well as to serve as a model for shaping technology *by many small but reflected steps* (compare also Rip et al. 1995, p. 8; Schwarz/Thompson 1990). The rationality reflection in this context enables technology policy to maintain long-term developments and plans though the method of proceeding is the incremental one allowing short-ranged flexibility requirements to be taken into account (for example, caused by social acceptance problems or societal learning effects). This model allows to chose an experimental approach in dealing with the challenges of shaping future. It enables learning processes at several stages.[7]

This model does, however, neither allow to talk about the plannability of society as a whole nor enable reliable predictions of future technology (Grunwald 2000b). The future is shaped by participants in the various practices of society, not by a central authority. Because the decisions are open to reflection and modification prediction only can deliver planning data but cannot anticipate the „real" development as it comes (Grunwald and Langenbach 1999).

Applied to the field of knowledge society this means that the transformation towards the knowledge society must not be reduced to an adaptation to a self-running development. Instead, it is a far-ranging challenge for society *to shape this transformation*. There are decisions to be made as selections of certain options (and exclusions of others) which are kinds of shaping processes even in the case that not all options may be taken into account. The question is not if shaping is possible at all but *how it will or should be undertaken*. This model of a directed incrementalism does not imply that there would is complete freedom in choosing certain options as is assumed sometimes in „shaping-euphorian" approaches. In contrary, there is no arbitrariness in selecting options: most options imaginable at all are not included into the set of options for selecting one by decision because of pre-decisions, pre-assessments, boundary conditions (Grunwald 2000b), ideas of acceptability and feasibility, the state of the art reached so far (Janich 1998b) etc.

In this sense there is no real option to prevent the knowledge society from coming to reality. We cannot decide if information technologies become more important in the future society or not. The challenge, however, is to shape *the way information technologies will be embedded in culture, in society and in the political systems*. At this level there are many options open for decision-making, regulation and also for technology assessment as reflecting such options *ex ante*. In the following this is exemplified by considering the field of internet, participation and democracy.

---

[7] Compare Otto Ulrich in this volume for learning at the level of regulation by the state, and section 5 of this paper for societal learning.

5 Technology Assessment and the „Internet-Based Democracy"

Participatory approaches to TA are discussed since the seventies and are, in the meantime, established in most European countries in different ways (consensus conferences, public discourses, mediation procedures for technology conflicts etc., compare Jean-Jacques Salomon in this volume). Shaping the knowledge society seems to be an appropriate field for such concepts because the transformation will change the lifeworlds of citizens (schools, shopping, banking etc.), the world of labour and the mechanisms of policymaking dramatically. This expectation serves as main argument for involving the citizens in the reflection processes on the transformation to the knowledge society. The way, however, by which such „stakeholders" are included in the processes of generating the „public opinion" and supporting decision-making, is not pre-determined by such a postulate. Participation at this general and abstract level seems to be a mere catchword: nearly nobody would oppose the call for participation. The real problems, in fact, arise by considering the details. This point is analysed in the following by looking at the vision of an „internet-based democracy".

The American candidate for president, Ross Perot, suggested the vision „People First". It consisted of an utopian model of a completely direct democracy based on the access to the internet for everyone. Citizens could vote directly and online at the occasion of each law under consideration. This method would be very cheap (after the implementation phase), very fast and repeatable on demand. The representative democracy then could be abandoned completely. Analysing this vision using the concept of pragmatic rationality starts with analysing the premises invested and the chances for justifying them. There are, at least, two main premises,

(1)   that the broad implementation and use of the technology „internet" itself will enable society to develop towards „more democracy"; the internet in this way is often described as a „democratic technology";
(2)   that forms of direct democracy will automatically increase the legitimation of political decisions.

Both premises, however, are untenable in this undifferentiated meaning. (1) What shall be meant by the formulation that the internet is a *democratic technology*? Firstly, this connection of words contradicts the rules for predicating at all: the attribute „democratic" may be used for states, procedures etc., but not for technology. Not a technology is more or less democratic but its *adaptation* and *enculturation* by society, its embedding into social networks (Mambrey 1998). Secondly, today there are about 40 millions internet users worldwide. That is a nearly vanishing minority. Most people even in the United States do not have access to the internet – a political system based on such a distribution would not be very democratic. Thirdly, talking about the internet as a democratic technology probably means that it is chaotic, has low hierarchy, that there is no instance censoring the inputs etc. These attributes, however, have nothing to do with democracy because democracy is characterised by creating legitimate decisions under the principles of majority on the one hand and the individual rights on the other (Sartori 1987; Waschkuhn 1998). Fourthly, there are, as a rule, different ways of embedding a certain technology in social affairs (the *enculturation* of technology, for example Mambrey 1998). It cannot be justified to assume one peculiar way of embedding as the only one possible – there might be other (and perhaps undemocratic) ways of embedding. In summary, for reasons coming from the social sciences, the philosophy of technology and the theory of democracy it cannot be justified to talk about the internet as a democratic technology – this would simply be misuse of language.

The underlying assumption that participation will *per se* increase the legitimation of political decisions, is highly problematic. The argument claimed to support this assumption – which is critically assessed in the following - is as follows: In general it is assumed that involving „stakeholders", persons or groups affected in decision-making will increase the acceptance of that decision because the acceptance behaviour, preferences, fears, emotions etc. of them then will be taken into account (Renn and Webler 1996). Thus, the legitimation of the ensuing decisions is expected to be improved. The internet is often seen as an ideal technology for realising such forms of participation (RTI 1996). The question is, to what extent this presupposition can be justified. The answer requires some reflection on the kernel of democracy and its major mechanisms to face the challenges deriving from new technologies.

The starting point is the observation that new technologies and innovations do not, as a rule, lead to pareto-optimal improvements of the former situation.[8] Instead, mostly there are winners as well as losers as a consequence of adopting the technology under consideration by society. This situation increases the demand for legitimation considerably: why should the potential losers – with respect to a particular political decision on technology – accept this decision though it would imply individual disadvantages? To create legitimation for such situations is the most difficult challenge for society and, in the case of success, one of its most important achievements: to create legitimation which demands people to accept decisions and their impact even in cases of individual disadvantage (Grunwald 2000a). There are democratically accepted procedures for handling such conflicts between the „common good" and the individual interests of people concerned. The crux is that the results of legitimated procedures – and this is part of the underlying cultural consensus on democracy – *must be accepted even if they are unwelcome.* Thus acceptance of the procedures means acceptability of the ensuing results. Procedures of this type are the means society uses to define the "common good" and to decide to what extent the public can be expected to accept a certain technological development in consideration of the interests of society as a whole – allowing for the fact that certain groups of the population may be burdened with the consequences of such decisions. If this would be no longer granted in a certain field, *the procedures would have to be changed* – a learning process in society has to be set up with the result of new or modified, again accepted procedures allowing for legitimate decisions at a new level. For example, this may occur, if legitimate decisions in technology policy lead to a dramatic

---

[8] Even if the pareto-optimal case would occur the question would be raised for the justice of the distribution of the benefits expected. The legitimation problems does arise, too.

rejection by society or if the decision-making procedures have to be modified for other reasons (for instance, in order to improve the chances for participation). The reflection on the pragmatic rationality and on the use of language should support such learning processes in the deliberative prephase of the modifications required.

What does this mean for the devise „people first" mentioned above? The wide use of the internet can enable the political system to perform internet polls to any problems under consideration and to get the results of those polls as fast as ever possible (nearly online). The internet thus seems to be a very suitable medium to accelerate polls and to listen to the „voice of people". But that has very little to do with democracy, the main challenge for democratic systems being *to generate legitimate decisions*. The legitimation of a political decision based on an internet poll, however, will be rather weak because the representativity of the participating users is not guaranteed. Who has participated, who has not, are there pressure groups determining the result by large engagement in the participation? Intransparencies or biases in such basic questions will prevent that decisions based on internet polls will be accepted as legitimate. Indeed, there are empirical experiences in the United States - having broader experience with the internet – that the use of the internet in the political system leads to new forms of mobilisation and lobbying instead to legitimate participation or an interactive democracy (Stegger 1996).[9]

The internet provides, in this context, many new possibilities of distributing information in a more quickly and very accommodate way.[10] Furthermore, there are new ways available for the self-organisation of stakeholders via networking and mailing lists. In this way, the internet seems to complete the set of media available for mobilising people, for information proliferation and for communication; however, it cannot replace the established mechanisms of generating legitimate decisions. Policymaking does not automatically become interactive if the internet is introduced. Why should citizens use the internet to participate if they are not using the traditional media of participation and communication? „Participation does not require technology but the political will" (Bernhardt and Ruhmann 1996, p125, translation A.G.). And, in addition, participation requires an appropriate cultural context and, if society is not prepared sufficiently for this aim, societal learning and active adoption of according decision-making procedures including participation. The internet my be used as a valuable instrument for certain supporting purposes but cannot replace such difficult learning processes. New information technology does, for the reasons mentioned, not automatically lead to more democracy. The way of embedding information technology into culture decides on whether democratic procedures are supported or hindered. In some countries with highly developed discursive tradition participate decision-making procedures are well established also in

---

[9] Even Bill Gates, one of the most important actors in this field, does not expect the necessity of legitimate political mechanisms to be abandoned in favour of an internet-based direct democracy (Gates 1993).

[10] However, it might be a fallacy and an unfulfilled expectation that the improved supply of information does automatically lead to an increased political engagement. There is no automatism leading from information to action (Stegger 1996, p 793).

absence of a wide use of the internet (take, for example, The Netherlands, Denmark and Switzerland).

If democratic principles and legitimation shall be maintained – and this is the case due to the self-understanding of our present culture – the state has to ensure by regulation the legitimation of the procedures of decision-making: the pre-conditions of the knowledge society must be ensured by political means (Mambrey 1998, p12).[11] An important aspect of this demand for regulation is the „universal access" (Miller 1996) as an indispensable element of using the internet as means of policymaking involving citizens. Otherwise the differentiation between the „information rich" and the „information poor" will undermine the fundaments of our societal configuration.

My counter-argument to the reconstructed premises of talking about the internet as a ‚democratic technology' is, in summary, as follows: firstly, the broad implementation of the internet does not automatically lead to more participation and, secondly, more participation does not automatically lead to more legitimate political decisions. It seems necessary to integrate participative elements, if required and supported by society, into legitimated procedures of decision-making. Representative democracy and direct participation are complementary, but they do not exclude each other (Waschkuhn 1998, p508ff.). To postulate participation always raises the question who are the persons participating in detail and why should the individuals which did not take the chance to participate (there is no obligation to participate) accept the vote produced by the participating part of the population (which may, perhaps, consist of a small group). The danger is that pressure groups and organisations will dominate such participation procedures, in particular if anonymous media like the internet are used.

For participative TA, this means that involving the stakeholders should not be restricted to taking into account simply their acceptance behaviour (Grunwald 2000a). In this sense, TA should set up *societal learning processes* with respect to the appropriate enculturation of information technology. Generating acceptance or transforming acceptance into decisions may be suitable for very special situations; but in many cases, TA must go far beyond such simple schemes. It should involve collective learning instead of merely mediating or compromising to attain technology acceptance.

In summary, the internet-based democracy has proven to be a vision without any rational and cultural fundament. The need for participation in affairs concerning new technologies should lead to a transformation of established decision-making mechanisms and a broadening of their perspectives but should maintain the achieved means of arriving at legitimate decisions.

---

[11] „Deregulierung und das Hoffen auf Selbstregulation wird dabei als Steuerungsinstrument sicherlich nicht ausreichen. Da informationelle Grundversorgung teil der Daseinsvorsorge ist, wird sie zur Aufgabe von staatlicher, also für alle verbindlicher Politik", (Mambrey 1998, p13), ... „wenn man eine demokratische Gestaltung der Gesellschaft will" (p15). „Ein hohes Ausmaß an Gesetzen, Normen und Konventionen ist erforderlich" (p15).

6 Conclusions

Rationality assessment starts analysing the language used by science, politics and the public. The language used to discuss about the knowledge society includes fears, hopes, expectations etc. The selection of the means of language for describing some aspects of the knowledge society is not independent of the normative sphere of values, positive or negative visions and the purposes to be reached by that selection (section 2, compare the contributions in Grin and Grunwald 2000). In order to achieve a transparent debate on the knowledge society very careful analytic work has to be done to uncover and to assess such connotations included in seemingly descriptive terminology. The assessment has to include at least three different fields of interest:

(1) the dimension of social sciences: what mechanisms of technology development are presupposed and how, or to what extent, can such presuppositions be justified (discussed in section 4 for the model of technology determinism);
(2) the dimension of ethics: what normative ideas are included in the terminology choosen and in what way can they, if ever, be justified from the moral point of view (for example, the dimension of privacy, data security etc.);
(3) the dimension of legitimation: what ideas about legitimate political decisions are transported by the terminology used, how do these ideas meet the societal consensus on what legitimation is required in order to ensure the acceptance of political decisions (exemplified for the relation of internet and democracy in section 5).

In summary, the approach of pragmatic rationality used in this paper opens the possibility to distinguish between more or less rational decisions relative to the cultural and societal „state of the art" (Janich 1998b, Grunwald 2000a). In particular, this leads to the concept of cultural adoption of technology and to the demand to consider the „enculturation" of technology relative to the standards of that society it will be applied to. Cultural pluralism, therefore, will not be overruled by global technology; global technology, in contrast, should be adopted differently in different cultural contexts.

References

RFTI 1995 Rat für Forschung, Technologie und Innovation (1995) Informationsgesellschaft. Chancen, Innovationen und Herausforderungen. Bonn
Basalla G (1988) The Evolution of Technology. Cambridge
Bernhardt U, Ruhmann I (1996) Revolution von oben – Der Weg in die Informationsgesellschaft. In: Tauss et al. 1996, p. 114-129
Bijker W, Law J (1994) (eds.) Shaping Technology Building Society.MIT Press
Bijker WE, Hughes TP, Pinch TJ (Hrsg) (1987) The Social Construction of Technological Systems. New Directions in the Sociology and History of Technological Systems, Cambridge (Mass.)/London
Gates W (1993) The Road Ahead. Penguin Viking New York

Gethmann CF (1996) Rationalität. In: Mittelstraß J (Hrsg) Enzyklopädie Philosophie und Wissenschaftstheorie, Volume 3. Metzeler, Stuttgart, pp468-481

Gethmann CF , Sander T (1999) Rechtfertigungsdiskurse. In: Grunwald A, Saupe S (eds.) Ethik in der Technikgestaltung. Heidelberg, Springer, pp117-152

Grin J, Grunwald A (2000, eds) Vision Assessment: Shaping Technology in 21th Century Society. Springer, Heidelberg

Grunwald A (1998) Technisches Handeln und seine Resultate. Prolegomena zu einer kulturalistischen Technikphilosophie. In: Hartmann D, Janich P (eds.) Die kulturalistische Wende. Frankfurt: Suhrkamp, p178 - 224

Grunwald A (1999a) (ed.) Rationale Technikfolgenbeurteilung. Konzeption und methodische Grundlagen, Springer, Heidelberg

Grunwald A (1999b) Rationale Gestaltung der technischen Zukunft. In Grunwald A (1999a, ed.), Heidelberg, Springer, pp29-54

Grunwald A (2000a) Technology Policy Between Long-Term Planning Requirements and Short-Ranged Acceptance Problems. New Challenges for Technology Assessment. In: Grin/Grunwald 1999 (in preparation)

Grunwald A (2000b) Handeln und Planen. Philosophische Planungstheorie als handlungstheoretische Rekonstruktion. Fink, München

Grunwald A (2000c) Rationality in Shaping Technology? In: Imre Hronszky et al. (eds.): Studies on the policy of Science, Technology and Environmental Issues. Budapest: MTA Szociologiai Intezete, (in press)

Grunwald A, Langenbach C (1999) Die Prognose von Technikfolgen. Methodische Grundlagen und Verfahren. In: Grunwald A (1999a, ed.), Springer, Heidelberg, pp. 93-131

Gutmann M, Hanekamp G (1999) Wissenschaftstheoretische Grundlagen rationaler Technikfolgenbeurteilung. In: Grunwald A (1999, ed.), pp. 55-91

Habermas J (1988) Theorie des kommunikativen Handelns. Frankfurt, Suhrkamp

Hartmann D, Janich P (1996) Methodischer Kulturalismus. In: Hartmann D, Janich P (Eds) Methodischer Kulturalismus. Zwischen Naturalismus und Postmoderne. Frankfurt: Suhrkamp

Janich P (1998a) Informationsbegriff und methodisch-kulturalistische Philosophie. Ethik und Sozialwissenschaften 9, Heft 2, S 169-181

Janich P (1998b) Die Struktur technischer Innovationen. In: Hartmann D, Janich P (eds.) (1998): Die kulturalistische Wende. Suhrkamp, Frankfurt/M., p. 129-177

Luhmann N (1990) Die Wissenschaft der Gesellschaft. Suhrkamp, Frankfurt

Mambrey P (1998) Perspektiven und Gefahren der Informationsgesellschaft. In: Ministerium für Arbeit, Soziales und Stadtentwicklung, Kultur und Sport des Landes NRW (Hg.): Neue Medienwelt – neue Lebenswelt? Düsseldorf, p. 6-17

Mambrey P, Tepper A (2000) Technology Assessment as Metaphor Assessment – Visions guiding the development of information and communications technologies. In: Grin/Grunwald (2000, eds.), pp 33-52

Miller S (1996) Civilizing Cyberspace. New York: ACM Press

Mittelstraß J (1998) Information oder Wissen – vollzieht sich ein Paradigmenwechsel? In: Bundesministerium für Bildung, Wissenschaft, Forschung und Technologie (Hg.): Zukunft Deutschlands in der Wissensgesellschaft (proceedings). Bonn, p. 11-17

Nennen HU, Garbe D (1996) (eds.) Das Expertendilemma. Zur Rolle wissenschaftlicher Gutachter in der öffentlichen Meinungsbildung. Springer, Heidelberg

Renn O, Webler Th (1996) Der kooperative Diskurs: Grundkonzeption und Fallbeispiel. Analyse&Kritik 18, S 175-207

Rescher N (1988) Rationality. Cambridge

Rip A, Misa T, Schot J (1995) (eds.) Managing Technology in Society. London

Ropohl G (1982) Kritik des technologischen Determinismus. In: Rapp F, Durbin PT (Hrsg) (1982) Technikphilosophie in der Diskussion. Braunschweig, p 3-18

Rüttgers J (1995) Bildungs- und forschungspolitische Schwerpunkte 1995. Bonn: BMBF

Sartori G (1987): The Theory of Democracy Revisited.Chatham New Jersey

Schwarz M (1992) Technology and Society: Dilemmas of the Technological Culture. In: Technology & Democracy. Proceedings from the 3rd European Congress on Technology Assessment. Copenhagen, p. 30-42

Schwarz M, Thompson M (1990) Divided We Stand. Hassocks: Harvester Wheatsheaf Press

Stegger M (1996) Partizipation und Demokratie im Cyberland – Politische Kommunikation im Zeitalter der Netze. In: Tauss et al. 1996, P. 785-801

Tauss J, Kollbeck J, Mönikes J (1996) Deutschlands Weg in die Informationsgesellschaft. Baden-Baden: Nomos

Waschkuhn A (1998) Demokratietheorien: politiktheoretische und ideengeschichtliche Grundzüge. München/Wien: Oldenbourg

Knowledge and Value Prerequisites of Evaluation and Decision-Making (Comments on Major Features of Knowledge Society)

Ladislav Tondl

1 Changes in the Goal Structures of Science and Technology Initiatives and Innovation Trends

Throughout the centuries, the key function of science and technology efforts was seen in expanding and improving the available pool of knowledge and the technical world, in broadening and enriching the environment which man is capable to control and whose advantages he can efficiently exploit. Only in the past few decades of this century has this image been substantially changing, showing that such endeavours have revealed not only their bright side but also their shadows, successes as well as risks, posing a broad spectrum of negative impacts. This has led many contemporary intellectuals to formulate countless new initiatives, pointing to what they called "limits to growth", issues of globalization and world-wide integration (the Club of Rome), the need to introduce a system-based and prognostically oriented assessment of the impacts of science and technology progress (TA), impose purposeful and knowledge-based constraints (Royal Society of Canada, the group of Nobel Prize laureates), and cultivate an "active", "cognitive" and "educated" civil society or a "knowledge society", and many other analogous initiatives. Also the author of this paper has been involved and often invited to take part in discussions on the impacts and trends of science and technology innovations, especially thanks to his study concerning what he described as a "double-faced technology", a work published more than thirty years ago, his accent on the great importance of primary decisions and start-up of some technology processes (Constraints to Freedom of Scholarship and Science, The Royal Society of Canada) and many other works devoted to the philosophy of technology and technological thinking and reasoning.

One can also hardly ignore attempts to deny the elementary function and goal-orientation of contemporary science and technology, attempts usually motivated by waves of irrational tendencies, fundamentalist ideologies, various forms of what is called "anti-science" or "alternative science", by what lies hidden behind appealing slogans promising easy solutions. But in actual fact, fruitful and effective solutions can only be achieved through an honest and time-consuming search in the intellectual spheres alone, i.e. in the field of science and technology as well as education, by engaging in a critical assessment of attained results and their potential impacts or risks. Also due to these reasons, current education, at all its age and intellectual levels, should be able to provide not only a body of knowledge concerning the possibilities of contemporary science and technology but also knowledge of the limits of its applications, its potential pitfalls, risks and inevitable restrictions. In addition to inculcating a broad spectrum of knowledge, the objectives of the required quality education include the necessary value equipment, which also encompasses ethical, aesthetic and cultural dimensions.

Stimulation of innovation initiatives in all walks of human activities, cultivation and improvement of their quality figure prominently among the social functions played by those intellectual spheres in the dissemination of whose results the existing forms of their transfer and all the useful information links should be taken into consideration. This also holds true of the human sciences, the social or humanities branches whose competencies should not be reduced solely to the social sphere but should rather be expanded to cover all the areas which are concerned, to a varying degree, with human activities or which result from human activities, i.e. all the areas involving actual as well as possible artefacts. Therefore, operating within such contexts, the social or humanities branches participate in humanizing those spheres, co-shaping what is sometimes characterized as the "human dimensions" or "human friendliness" of all the sectors dealing not only with man's actual life but also with the prospects of the human race and their potential risks.

The last comment also implies that certain value-related structures tend to enter into the goal orientations of all the departments of knowledge-based activities mentioned above, that an element of individual as well as social responsibility cannot be ignored in those activities. Basically, however, these are now considerably changed value structures, differing - in many respects - from the value systems held up by the advanced industrial and consumer society, particularly the values of material consumption, the cult of growth, material wealth, glorification of size, of technical and economic perfection, efficiency, i.e. optimization of the input and output ratio in activities viewed in an isolated perspective or a solution of a similarly conceived problem situation.

Reflecting on what is sometimes called science's "social function", and - in parallel to that - also on the social function of the highest levels of education, we have to realize that not only knowledge itself but also value-related contents of such functions, those relating to the actual research and educational requirements, to the actual needs and expectations have been changing considerably, that new requirements, new demands and new expectations have been emerging at the end of our century, also in connection with an outlook covering prospects for the next century and the next millennium. Such new requirements and expectations arise together with social, cultural and value-related trends in the development of a society marked by various attributes, notably terms such as "post-industrial", "post-modern", or rather with processes classified as "globalization", "integration", construction of an "information society", "educative society", "knowledge society" and some other terms.

It is quite natural that the goal orientation of science, research projects and the highest levels of education has its own traditional constants, including expansion, improvement, and diffusion of the existing body of knowledge, factors known to promote what is generally called society's available intellectual wealth. However, the current approach to this intellectual wealth, surviving from the industrial society, lays greatest stress on the immediate application of that fund, i.e. in material production, its growth and, hence, rising consumer demands. This is associated with the persisting and frequently highlighted socio-political and economic priorities, i.e. the cult of size, growth and vastness, with an admiration for the scope and bulkiness of technical projects, their efficiency and range of capacities. In parallel, the modern era has witnessed a continued upsurge, although accompanied by mounting protests, of what the author of this paper has called the "cult of gradation", an admiration for size and all the properties summed up in the word "superlative". There is a gradually expanding gamut of areas and activities connected with them whose values of self-imposed and conscious restrictions are given prominence (such activities include, for instance, smoking or drinking alcohol), while normatively conceived restrictions, bans and sanctions connected with them are being introduced in other fields. Some thinkers believe that the values of asceticism, qualities so highly appreciated in the Middle Ages, tend to reappear at the end of the century and the millennium. The past few decades have also spotlighted the importance of values whose history is actually quite old and whose significance grew with the downfall of feudalism and absolutism, i.e. the packages of civil rights and liberties whose revival has made a sizeable contribution to the downfall of latter-day brands of absolutism and 20th century totalitarian systems. All the signs are, however, that these values continue to be threatened and violated. Efforts to revive the topical character of human rights have been accompanied by the emergence of an extensive group of issues, which can be generally characterized as focusing on the respect of the rights of nature. This is not only because some natural resources are not endless and inexhaustible, this particular category including - besides energy resources - also water and air, but also because many cases of human interference with the rights of nature are irreversible, cannot be corrected, and pose lasting damage, thus actually jeopardizing the future of the human race.

These and many other changes in existing sets of values, which are required to be respected in all walks of human activities, substantially affect the overall goal orientations of research and education. Whereas, up to the middle of this century, achievements in science and the results of its applications were mostly used to flaunt the possibilities of transformation, of man's active intervention into the natural and social environment (including faith in the possibilities of social engineering), changes in the existing sets of values are now highlighting the importance of preservation, maintenance and also the significance of protection against potential threats, prevention of possible risks. The requirement of "sustainability" has been raised, while the actual spectrum of what could or should be sustainable is being steadily expanded. While the nineteenth century, if one may borrow Marx's well-known tenet, wanted not only to explain but also to transform, the latter half of the twentieth century, and especially its end, is posing more vigorously the demand to preserve, maintain, protect and - on many occasions - to salvage what has managed to survive thus far. Basically, these are not totally new requirements. Analogous principles have always figured among the traditional values of the prevailing way of thinking and reasoning in medicine where learning, i.e. a series of steps leading to a diagnosis, has been associated with measures conducive to the preservation and maintenance of health, and - in many cases - life as well.

2   Human Dimensions and Human Values

The term "human dimensions" has appeared in the development of science - and especially in the development of the technical applications of acquired findings - as something which can be overcome by those applications, which can be surmounted several times over. Originally, people used their own measures, such as a foot, an ell, the length of a step or the strength of their beasts of draught ("horse power" for instance), as an essential yardstick to measure what went on in their world. New problems of the human dimensions which have, up till now, formed the core of the themes of several major disciplines, notably ergonomics, problem areas called "social mastering of technology", "man-machine relations" and some other branches, have emerged in the wake of growing capacities, the technical and economic potential of machine systems, manufacturing systems and production lines, efficient exploitation of the existing complex technical facilities. This has been instrumental in raising issues as well as doubts concerning man's limited powers and abilities, concerning the possibility of replacing some human activities by automated systems, the roles and problem situations for the solution of which man's physical and intellectual capacities have proved to be inadequate, and thus questions relating to the limitations and insufficiency of some human dimensions. Somewhat different fields of issues, doubts and disputes relate to the concept of man as the centre of the world as well as an absolute yardstick of anything that happens around us, i.e. a concept usually referred to as anthropocentrism. This involves the traditional dispute known since Antiquity, whose extreme position is the maxim that man is an unlimited lord and master of the universe who, in his mastery and exploitation of nature, need not restrict himself by anything. Opponents of this extreme brand of anthropocentrism justifiably counter by saying that man is an integral part of live nature, that he should respect its rights or - as they point out - that man and nature are both subjected to the principles or laws of a transcendental nature. Whatever views we may hold or whatever sympathies we may have for different standpoints in such disputes, we have to take into consideration the fact that both the spheres of human dimensions and their applicability have their own justified limits.

Problems concerning human dimensions have also been encountered in yet another sense in the relations between elements of the technical world on the one hand and man on the other. While, at the dawn of the industrial era, elements of the technical world and technological facilities generally appeared to be a multiplication of human forces, first physical and gradually intellectual as well, figuring as obedient and loyal servants who, true to say, may sometimes lead to an accident but who are, essentially, prepared to be subordinated faithfully to man and serve him as a "trestle" - to borrow a term from Martin Heidegger -, it soon became evident that, in a way, these facilities tend to control man, forcing him into a certain inevitable behaviour pattern (as used to be the case in some production lines). In a sense, man has changed, within such contexts, from a subject who could freely manipulate with this helper or "trestle" (Gestell), into a subordinate who had to adjust himself to technological rules, cycles or rhythms.

The term "human dimensions" has appeared in other new contexts in technological thinking and decision-making in projects aimed at providing for easier and more acceptable forms of handling complex technological equipment. This concerned primarily projects involving the highest generation of computers and information technologies. Of great importance in this respect were efforts to eliminate complex and relatively challenging links between a computer user and the computer itself, to introduce voice control, to use direct dialogue. In these contexts, the "human dimensions" have been perceived as modes of control, manipulation and communication which are closer to customs and habits formed over generations. There emerged requirements for technical facilities to respect man's inherent and, in a sense, humanely natural modes of behaviour, reactions, their time dimensions etc. That was why the concepts designed to control and utilize those facilities have laid greater accent on direct communication or dialogue, that was why further progress has been made to improve the possibilities of communication in natural languages, and the use of pictorial or graphic modes of expressions has been increased. It was precisely within these contexts that the need of enriching the principle of "human dimensions" or "human friendliness" with new contents has been spelt out.

The issues of human dimensions are, however, much broader and more general. Therefore, they can hardly be confined solely to situations whereby we sit down at a steering wheel of a car, simply start and run a machine, use a computer or a terminal of a computer network. By taking these or analogous steps, we actually enter the realm of human artefacts, a world that is sometimes called a "second nature". But the question is: does the proper, original nature, untouched by man, still exist? Are we not actually destroying many elements of the second nature just because these have lost their original functions, just because they have become outdated, just because they have been replaced by us with new, more prefect and more efficient artefacts? Is it still possible to preserve the original nature unspoiled by man? That cultured or sophisticated communities have not been and still are not indifferent to these and other similar and analogous questions is corroborated by their diverse attempts or endeavours to build up art collections, museums, open-air museums, protected areas, reservations and other islands of what has to be preserved, what bears the hallmark of human values as well as results of what has, up till now, been viewed as manifestations of cultural, aesthetic and also ethical value-related attitudes. (They have been known to organize such collections virtually since the time of Renaissance). Similarly, attempts and efforts, which also represent highly specific artefacts in their own right, can be seen as demonstrating the human dimensions and human values. But not even in those contexts can we avoid answering the questions which quite naturally arise with a good deal of pervasiveness: Should such manifestations of humans dimensions and human values be saved and preserved only in protected open-air museums and reservations?

Seen in this light, the subject of human dimensions and the application of such standpoints or criteria derived from the way of respecting the individual aspects of human dimensions is neither unequivocal nor easily interpretable. In an effort to apply these positions and criteria, especially in decision-making and evaluating roles, during processes seeking to apply new findings, discoveries or technical solutions, two major fields of relevant circumstances should be taken into account:

1. First and foremost, we should realize that the groups of aspects or criteria, collectively called the "human dimensions", are considerably non-homogeneous, containing highly divergent aspects concerning somatic, psychic, cultural and moral factors whose importance may greatly differ in various decision-making processes or evaluating roles. Hence, the process of accentuating "human dimensions" has, primarily, the character of stimulation or motivation, aimed at the nature of what is substantial in the given context to be further specified.

2. There is another field of additional circumstances and contexts which is of great importance for the application of a selected package of criteria proceeding from various aspects of human dimensions and also for the selection of the relevant criteria. These include respect for the nature of the task being solved, selection of applicable instruments, scope of all types of potential risks and anticipated impacts, respect for time and spatial contexts etc. When deciding, for instance, about the start-up of a technical solution, much depends on whether such a solution should be located close to a densely populated area or in a locality where threats to human population are either non-existent or at least minimized. Since people have not yet succeeded - and are not showing any signs of succeeding in the foreseeable future - in ruling out some potential risks in many areas of applications of new scientific findings and new technological solutions, or rather, some newly designed therapies (not only in a medical but also broader social sense) we are justified to consider several possible alternative solutions. The following alternatives rank among the most important ones:

   - a total ban on those applications, methods, processes or solutions of the given problem situation whose potential risk, possible threats or long-term negative impacts exceed an acceptable limit (this is - or should be - the case of nuclear test explosions);
   - fundamental restrictions imposed on such applications to a limited extent, a limited number, a limited locality, usually while applying strict control or otherwise defined specialist supervision;
   - concentration of such procedures, applications or experimental verifications to limited and consistently controlled areas. (This last of the given alternatives is sometimes described as the "dead is dead strategy". It is, however, evident that for the determination of those localities and areas, for instance in great depths beneath the Earth surface, in regions in outer space distant from the Earth or in other cases, we never have clear-cut guarantees and, consequently, potential although still unknown risks cannot be excluded.)

3 Changes in the Value Structures and the Vision of a "Knowledge Society"

The accelerating pace of innovation steps, which has continued since scientific discoveries and their technological applications connected with progress in the branches of physics and the opening sphere of biotechnology as well as information technologies, has not always been accompanied by desirable changes in the sphere of value structures, in the field of life goals, and also in the corresponding areas of ethics. The hitherto dominating value structure is strongly dependent on what used to form the objectives, ambitions and preferences originating during the heydays of the advanced industrial society. If it is true that its upswing has led to widening the gap, seen from the viewpoint of the values of the industrial society, between the most advanced countries and developing nations or - to put it differently - between "North" and "South", then an equally significant delay may be traced in the sphere of knowledge and values. Nothing can be changed in that conclusion by the realization that since then the sets of values cherished by technologically and industrially most advanced countries have also spread from the centres in Europe and North America to some Asian countries, that those Asian states have also been grappling with some of the problems, difficulties and troubles caused by the industrial, consumer and - to a certain extent - hedonistic and greedy society. It is, however, only natural that these particular problems and the darker aspects of human lives connected with them should not be allowed to cast doubts on the momentous accomplishments which have resulted in eliminating many threats of the past, as shown by the eradication of the formerly frequent epidemics, prolongation of human life, alleviation of former famines, reduction in child mortality rate, the opening of not only transport but also information routes, and a number of other phenomena the past generations could only regard as wishful thinking.

If the development of applications of science and technology in the twentieth century, and especially in the past few decades of this century, constitutes a rich spectrum of positive as well as negative factors, bright sides as well as shadows, achievements as well as risks, including actual and possible breakdowns and accidents, then these relatively new phenomena and the positions proceeding from them are matched by an equally variegated spectrum of value structures. It is, therefore, possible to reconstruct groups of certain types of counter-positions or contradictory values, which often contravene one another and project themselves into a pattern of detected, discussed or analyzed problem situations. We will now try to outline some of the major counter-positions in contemporary thinking, decision-making and assessments related to the results of science and technology innovations.

1. According to the traditional values of modern science and technology, reinforced by the rediscovery of the values of Antiquity, rational thinking and decision-making and re-affirmed by the flourishing of the hitherto acquired knowledge and possibilities of its application, man, perceived as a rational being, is the master of nature. His knowledge accords him power, as spelt out by Bacon's well-known tenet likening learning to power (scientia est potentia).

This power, in turn, provides an ever-expanding space for control, transformation, manipulation, including manipulation with nature's resources and capacities, whether these are material, energy or intellectual. This is often accompanied by an illusion that these resources and capacities are so vast that in their practical applications we need not take their finite or limited nature too seriously.

Numerous objections have been raised against the different aspects of this particular approach to values: Man is not only a carrier and user of knowledge about the world itself or - if we can use the renowned symbolic expression of Kant's epitaph - a subject of the knowledge of the starry sky but also of the knowledge of the moral law, a subject and hence an initiator of responsibility, a protector of human as well as natural values. This includes, above all, protection against possible abuse of new results of science and technology development, protection against their adverse impacts and potential risks. If an overestimation of power stemming from knowledge, an overestimation of the extent of this space man is capable of mastering, tends to create feelings of self-confidence, power or strength derived from mastering and exploiting natural resources and man-made technical artefacts, then the need of protecting oneself and others against possible negative impacts, an awareness of potential dangers or risks justifies the existence of a gamut of other ethical attitudes, namely feelings of responsibility, modesty and humility, an awareness of the limitations or insufficiency of the hitherto acquired knowledge, of the need of imposing certain limits and restrictions. These tendencies and gradually asserted changes in values also tend to stimulate a new "search", if we can use in these contexts H. Simon's general description of cognitive efforts.

2. At the apex of its system of values or its system of priorities the industrial society has placed those values which were associated with power, control, exploitation or consumerism, giving preferences to material goods instead of intellectual and spiritual qualities, which were connected with a system of activities leading up to those value systems. It should be admitted that these value orientations have, at least occasionally, been compensated for by manifestations of altruism, charity, patronage but these activities have also been displays of power, control, ostentatious shows of material riches. Demanding solely what is generally characterized as postulates of "redistribution", the critics of those value orientations actually proceeded from the self-same value-related attitudes.

However, opposition to the values mentioned above and to the attitudes proceeding from them, to the goal orientation of different activities is - in no case - solely a matter of modern times. Already the principle of the division of power, efforts for decentralization of various forms of control, different kinds of restraints to consumerism (whose classic examples include fasts, self-imposed asceticism etc.) have long-standing traditions. The development of medical knowledge, of healthy lifestyle, of sound nutrition as well as other departments of knowledge has given rise to whole new sets of stimuli for justified and voluntary restrictions in diverse areas of nutrition, starting with the awareness of the harmfulness of drugs, smoking, consumption of foods with an unwholesome biochemical composition etc. It was already during the development of the consumer society that various warnings, appeals for voluntary restrictions and naturally also justified forms of prohibitions penetrated into its hierarchy of values. But in view of the current level of knowledge concerning possible negative impacts of various steps in life or situations, which are still viewed as normal or quite acceptable, the scope of those restrictions probably accounts for just a portion of what man should really keep away from. Whereas a civil society respecting its human and civil rights and liberties is still regarded as the key goal of the future orientation of society's development, the term "civic society" itself has been given other attributes, i.e. educated, well-informed and responsible or - as the same attribute is sometimes succinctly summed up - "knowledge society".

3. A typical feature of the value structures of the industrial and consumer society was the cult of change, a high regard for growth, size, degree of efficiency, sequence of changes which the author of this paper has called the "cult of gradation". In similar contexts, some other thinkers speak of a high social prestige given to megalomania, which was, however, viewed and also largely interpreted in the spirit of the anthropocentric concept of the world and especially the world of artefacts, i.e. second nature. But each of the given terms has its own specific features, while particular illusions have developed around some of those terms. Suffice, for instance, to mention the well-known illusions concerning a path leading towards greater and steadily growing progress, illusions used by one of the ill-famed ideologies as a path towards a "brighter future". The traditional concept of efficiency or cost-effectiveness of technical changes or any investments is confined to an input-output ratio conceived in an isolated fashion or rather to a cost-benefit ratio without considering a wider range of potential impacts in broader temporal and spatial contexts, regardless of the fact that an eventual reduction in the level of entropy in a given area under scrutiny or in a given sphere of interest will ultimately lead, in the immediate or more distant future, to an increased level of entropy in a more generally conceived environment. The same applies to the highly controversial but greatly appreciated growth-rate or cult of gradation.

As early as in the 1940s, Norbert Wiener, the founder of cybernetics, emphasized that what we call science and technology progress is not achieved for free, that the attained changes always have to be paid for, while the exacted price may even be disproportionately high. Last but not least, also many substantial changes instrumental in creating material, energy as well as information processes typical of the growing industrialization have posed a rapid threat to various spheres of the environment, leading to a deterioration in the quality of air and water, to the so-called greenhouse effect, to desertification, upsetting many originally well-balanced states and possibilities of self-reproduction, threatening certain protective mechanisms, such as the ozone layer etc. Seen in this light, it is by no means accidental and unfounded that some leading thinkers have been speaking of a "stress of our planet" or a whole series of what are today absolutely real and confirmed risks.

The emergence of these and similar situations has given rise to new goal orientations in various walks of human activities. These do not concern solely the application of a traditionally conceived rationality but a much more general - or as some authors put it - global or humanistic rationality. In addition to objectives associated with desirable or necessary changes, a system of goals and values connected with preservation, maintenance, with efforts for the sustainability of those states and values which are likely to secure continued and acceptable future of the human race has been established. Instead of formulating the requirement of continued economic, technical or otherwise quantified growth, authors wrote studies describing the limits to growth, the limited nature of resources and capacities, the tolerability or acceptability of some already attainable performances. (One of the first ground-breaking initiatives in this context was formulated by a group of the members of the world-known Club of Rome. In a similar vein, numerous environmental initiatives were launched in different parts of the world). One can, therefore, claim that in many rational social activities a traditional optimistic spirit has now given way to that of maximum criticism, to requirements of careful verification or confirmation, greater carefulness and responsibility.

If, indeed, a new value situation has emerged at the end of our century, its salient feature is that it no longer concerns solely specialists, scientists, technicians and top decision-makers but is of interest and concern to ever larger groups of citizens. If, indeed, there are changing goals in some departments of learning, which are also known to be studying potential threats, risks and also ways of preventing and fighting the adverse effects of various human inroads made into the existing natural and social conditions, including social relations, it is only desirable for a civil society to be formed by active citizens and their communities who are well-informed and who also demand to be given desirable information, who are capable of voicing their apprehensions and initiatives in a matter-of-fact and cultured fashion. This requirement also applies to the need of introducing sophisticated forms of dialogues between experts and citizens or civil groups and initiatives. The term "knowledge society" (or terms with basically the same or very similar meanings, such as "educative society", "educated", "cultivated" civil society) has been coined for a civil society viewed in this particular way. Such a society should have a naturally complex, preferably multi-layered, structure vital for asserting personal, group, regional and other partial interests, which may find themselves in different relationships to the existing globalization trends but which can never afford to ignore them totally. The vision of such a society should respect a number of requirements of which the following are particularly significant:

- Knowledge society is, first and foremost, a civil society, i.e. a democratic, open and thus pluralistic society of free citizens who are entitled to be adequately and reliably informed, a society which, however, has at its disposal ways and means of protecting itself against any abuse of freedom, restrictions and threats to its civil rights and liberties.
- Such a society is known fully to support such resources and networks facilitating access to all forms of information and consequently to knowledge, while never ignoring the prominent role played by the recipient of learning, his or her knowledge and value equipment, his or her ability to accept, assess and select necessary, desirable and, in a given situation, relevant data and knowledge.
- Needless to say, not all the members of such a society are expected to share the same prerequisites for understanding and mastering all the knowledge; such prerequisites are always spread unevenly in any human community. Still, a

substantial section of the citizenry is capable of respecting all the values of learning, the importance of education, maintaining the necessary trust in verified and adequately confirmed knowledge and also in the carriers of such knowledge.

- Also those members of society who possess adequate knowledge or rather who share in its generation and dissemination, in the educational processes and all forms of dissemination of the results of science and technology should aim at facilitating access to such knowledge, its comprehensibility or transparency, while pointing out, in a responsible way, all the positive aspects as well as eventual risks associated with those results, thus participating in shaping a responsible and critical civil attitude.

- The necessary communication links between creative members of the basic intellectual spheres, i.e. links sometimes characterized as a dialogue between science and society, a dialogue between the cultural sectors and society, have the nature of a dialogue conducted between equal and full-fledged participants in a discourse who are known to share respect for their values, who sensitively listen to the subjects proposed by the other side and especially to mutual requirements among which innovation initiatives occupy a prominent position.

One may argue that the given requirements placed on the knowledge society constitute more or less an ideal type (in the sense of sociological methodology) rather than describing an actual state. Such an objection may only be countered by claiming that in all the previous development stages society's leading intellectuals have always led the way in formulating certain new images, ideal models, visions and often mere utopias, stressing that societies of the day should seek to emulate or at least approach such images or visions. As for our own socio-political and economic situation, i.e. the situation of a society which has passed through certain stages of social, economic and - predominantly only - institutional transformations, changes that have, furthermore, created the illusion that they are in a position to exercise a more substantial influence on the patterns of behaviour, value structures or preferences established for two generations of totalitarian and monopoly power, certain corrections of shortcomings appearing in the stages of transformation processes so far are, undoubtedly, urgently needed. Wholehearted efforts to restore the prominent position of the values of learning and knowledge, education as well as intellectual and cultural heritage in general may be part and parcel of such corrections, although they should definitely not be seen as a panacea.

4 The Necessity to Introduce an Efficient Network of Communication Links

Contemporary research centres and institutions of the highest levels of education have long ceased to be crammed into the narrow pigeon-holes of cities, regions, nations and states. Thanks to present-day communications, information technologies in particular, the world has grown smaller, and all the major institutions engaged in cultural, intellectual and cognitive activities and operating with their results are concerned, in one way or another, with much broader dimensions, and usually have a world-wide framework, at least in science and in the key sectors of culture. This also reflects the prevailing integration and globalization tendencies mentioned above. The current trends in globalization and "the world's diminishing dimensions" in the sphere of research, development and educational activities are known to be closely related to several problem areas of which the following should be singled out as particularly important:

- the field of communication, including communication forms and instruments transcending the confines of a single workplace, region and state;
- the establishment of well-organized networks, including networks facilitating mutual control and criticism;
- the sphere of co-ordinated contacts and competition, complete with measures to safeguard competitive skills and prerequisites.

Due to their nature, *research activities* and all forms of *educational* processes represent a form of *communication*. Scientific learning is sometimes characterized as a researcher's communication or dialogue with nature, with social milieu and, consequently, with the hitherto acknowledged or attained level of knowledge, with what has been preserved so far thanks to established social memory. Direct or suitably mediated information contacts and transfer of information interconnecting all those studying the same or similar subjects, without any limitations of national frontiers or language barriers whatsoever, are absolutely vital for the contemporary level of science and research. The present-day information technologies, including direct contacts mediated through the Internet-type technologies, are in a position to secure such direct contacts. Moreover, there are also many other forms of suitable channels of communication, some of which boast of long-standing traditions. (These include what are known as "invisible colleges", i.e. exchange of information among specialists working on the same subject, who communicate especially by exchanging manuscripts of their works before publication, i.e. so called "pre-prints", or attending working meetings devoted to selected thematic fields etc.)

No less important is direct communication between individual departments or faculties at universities on the one hand and corresponding institutions at domestic and foreign universities on the other. Some of these contacts tend to develop spontaneously, while firm organizational frameworks are being established for many other such links. As for the European Union, German and French universities situated in the Rhine region are known to have accumulated very good experiences in this field. This applies not only to the exchange of teachers but also students, recognition of exams and degrees etc. The term "visiting professor" has become a household word in the English-speaking world, denoting a matter-of-course and fully recognized need of maintaining lively contacts and good standards of the university concerned. Another noteworthy phenomenon is well-organized cooperation promoted in an effort to attain highest possible quality (described by the well-established term QA or "quality assurance"). Appropriate databases are set up as part of such efforts, while agencies operating with these databases keep their members well informed about latest developments etc. In practical terms, this means the formation and application of purposefully created *information networks* which, operating from established centres and available databases, facilitate the performance of many intellectual activities by warning their users against "blind alleys" of those cognitive processes called by H. Simon a search in the maze and described by J.-J. Salomon as an "uncertain quest". Needless to add perhaps, it is vital to warn against an uncritical overestimation of the power of such instruments: Only scholars who already possess a high competency may benefit from using such information networks, which can hardly be expected to make up for talent, creative invention, tenacity and patience, properties required by genuine creative work.

A trouble-free and efficient functioning of information networks is particularly important for strengthening and maintaining international projects, for joint programmes in relation to those research centres which may be singled out as centres of excellence, i.e. outstanding and stimulating workplaces capable of blazing new paths in science, formulating new and stimulating initiatives and also warning against previously unknown dangers. Supranational research projects in particular, joint programmes involving top-level preparation of highly qualified specialists at today's level are virtually unthinkable without fast and reliable communication channels, for instance of the type of electronic mail. As is well known, speedy and efficient exchange of information is vital for the research of some parallel projects, for instance spread of certain contagious diseases or infections whose scope transcends the boundaries of a single region or state. This applies, to a large extent, also to the transformation processes, their typical features, risks and their possible errors in the post-communist countries, nations which have actually been hit by a similar infection of their totalitarian regimes and ideologies.

The need to promote efficient communication links and information networks in the academic sector has become ever more topical in the wake of the newly emerging trends in areas which may be characterized as securing an entry of the *competitive spirit* and competition into those sectors. At present, a number of top-ranking posts, especially those of directors of institutes, section heads and other posts, are filled by winners of competitions of tenders. Also academic posts are filled by selecting the best applicants from a large number of candidates. The principle of competitiveness has been successfully applied in the systems of awarding grants as well. This makes it possible to weed out impracticable research ambitions. Figuratively speaking, it is necessary to eliminate any attempts at "discovering perpetuum mobile" and also "discovering America". The introduction of a competitive spirit and competition into academic life has also laid greater claims on the level of information and equally on the overall quality and competence of appropriate teams of experts, evaluating and decision-making panels. As a result, a system of dual, follow-up evaluation is always in place: assessment of assessors, covering not only their qualification and professional erudition but also their civil and moral records and responsibility as well as an evaluation of applicants, candidates, project authors etc. At the same time, both levels of evaluation should also be focused on the past as well the future, covering the available capacities, time horizons and available conditions.

All the forms of communication processes, especially well-functioning contacts with the most outstanding centres in the whole academic sector, can thus contribute to stimulating innovation initiatives and, simultaneously, creating a critical climate marked by an awareness of one's responsibility for the future of the given society. At the same time, it is true that the young generation, preparing for its highly challenging and qualified activities, will have to undergo permanent training, i.e. expand its knowledge as well as scope of responsibility in an accelerating rhythm of knowledge, social, technical and economic changes and possibilities of new risks or other dangers brought about by those changes. A growing differentiation within the systems of research centres and universities, within the systems of goal orientation of these institutions ought to be accompanied by a higher level of mutual information links. The past decades have shown that isolation in terms of information, isolation in the broadest sense of the word, has led to a loss of talent, creativity and innovation initiatives, that such an isolation has caused damages fully comparable with the losses ensuing from any restrictions of the freedoms and autonomy of major intellectual centres as well as from any power, political and ideological interference into their own activities.

Ethical Norms for the Selection and Transfer of Knowledge

Stefan Berndes

> "Es gibt einen Grad von Schlaflosigkeit, von Wiederkäuen, von historischem Sinne, bei dem das Lebendige zu Schaden kommt, und zuletzt zu Grunde geht, sei es nun ein Mensch oder ein Volk oder eine Cultur."
> Aus: Nietzsche, F.: Unzeitgemäße Betrachtungen II, Vom Nutzen und Nachteil der Historie für das Leben.

1 Introduction

The characteristics of some projects or systems seem to demand the transfer of knowledge in order to be able to handle and use them, and to minimize their negative long-term effects responsibly. If we accept this, the question for ethical norms of selection and transfer of knowledge becomes relevant. And knowledge grows. This thesis is accepted by many economists of science, knowledge sociologists and philosophers of science. If endless growth is impossible in a world of limited resources, and if one needs resources in order to preserve knowledge, than it could well become necessary to selectively preserve knowledge in the future. In this situation, I would like to propose to discuss apart from economic criteria of knowledge selection and transference ethical ones.

The aim of this presentation is to motivate norms of knowledge selection and transfer and to sketch justifications for them. In doing so I would like to give you some ideas of my understanding of key concepts such as knowledge, knowledge selection and transfer, and oblivion of knowledge. With the help of some examples I would like to stimulate interest in the topic before I proceed to outline a catalogue of norms. This needs some explanations about metaethical premises, the argumentative process, and I would like to list the arguments which have been selected for the justification of the norms.

Then I would like to counter the objection that ethical norms of knowledge selection and transfer in a longterm perspective are nothing else than a reflex of a certain basic conservative attitude. In my summary I will end up with some prospects for fields of application and further questions of research.

2   Knowledge – Oblivion, Selection and Transfer

It was Plato who defined knowledge as justified true belief.[1] In contrast to his understanding of truth as being co-extensive with being, the mainstream of modern epistemology describes the holistic and coherentistic character of justification. Therefore justified beliefs are true, and there is no certain and revision-resistent knowledge.[2]

Knowledge of a single person could be defined as their individual system of beliefs. It consists of propositions of different certainty (Gewissheit) and centrality. From this system the person is able to generate justifications for further opinions.[3] I understand knowledge of a society as the system of beliefs which is generally accepted in such society.

Knowledge consists of theories. Theories combine propositions which before seemed to be independent.[4] Presumably they neither result from observation data alone nor are they axioms to be finally founded (Letztbegründungsversuche) successfully. Axioms become generally justified by the performance of the theory (Systematisierungsleistung) itself.[5]

In this view scientific knowledge cannot be distinguished from common knowledge (Alltagswissen) by its certainty. But one can distinguish it by its resistance against (institutionally protected) claims of justification and its generation based on methods.[6] Scientific organisations[7] support the coordination of research and education (Lehre und Forschung). They stabilize the actions of the scientists which are lead or motivated by the scientific organisations.

Knowledge is always transferred by information. It is understood information which has been integrated into the context of already existing knowledge.[8]

Is oblivion of knowledge as an action[9] possible? To explain oblivion – in the case of scientific knowledge – is easy using the previously presented holistic and coherentistic concept of knowledge: In the course of changes or alterations of theories valid propositions get lost. The loss of information which is needed for a competent reader to reconstruct knowledge could be caused by the loss of texts.[10]

---

[1] In Platos Theaitetos, 200d - 201c this definition is considered.

[2] For a brief presentation of this concept of knowledge see: (Nida-Rümelin 1996) p. 54-63.

[3] Comp. ibid. p. 42.

[4] Theories combine propositions which before seemed to be independend ("Theorien verknüpfen Propositionen [...], die zuvor unabhängig voneinander erschienen."); comp. ibid. p. 39.

[5] Comp. ibid. p. 45.

[6] For the definition of science comp. (Mittelstraß 1996) pp. 717. Institutions determine our action as types of this action in specific situations for everybody even if action does not follow this institutions in singular cases by e.g. sanctions or desires; comp. (Schwemmer 1984) p. 250. I understand science as institution sense-oriented (sinnorientiert). I.e. scientists are motivated by scientific institutions to act according to them if they understand them.

[7] Organisations are social systems of a special kind, which produce certain results by the motivation and coordination of actions which can be expected only from members of it; comp. (Luhmann 1984) column 1327.

[8] Comp. (Kornwachs 1998) p. 8.

[9] Action is a behavior which one could have refrained; comp. (Kutschera 1999) p. 337.

[10] Kornwachs differs between a lost of texts due to a lost of data storages and a lost due to oblivion of read-write-technologies; comp. (Kornwachs 1999b) pp. 7.

If artefacts which are necessary for scientific or technical knowledge are destructed or are no longer available, some effects cannot be produced, observed or measured any longer. Then, the meaning of scientific texts is not guaranteed and shifts in their interpretation are probable. Oblivion could also be an effect of changes in the scientific methods or altered interests.

The selection and transfer of knowledge could be understood as a collective action in uninterrupted chains of tradition. Organisations assumed responsibility for this task. Of course, it is possible to transfer semantically stable[11] selected knowledge from one organisation to another.

Knowledge is inferable from messages which have a limited lifespan. Therefore longterm transfer of knowledge needs intelligent copy processes.[12]

Decontextualisation of knowledge is possible even in institutional continuity because the alteration of theory, methods and interests cannot be stopped. If such changes become visible one should pass on the contexts using a correspondence principle. I.e. "advanced" contexts of the knowledge should include contexts as special cases which are still relevant.

3   Problems of Knowledge Selection and Transfer

Up till now we have identified three classes of knowledge selection and transfer problems in which one should take into account ethical criteria for the decision too:

1. Sometimes ethical arguments are already in the debate as to whether or not we are obliged to inform about our longliving maybe dangerous estates.
2. The argument used in the first class is as follows: If we have started an "intertemporal cooperation project"[13] then we should explain and inform future generations about such project. This seems to be – at first sight – uncontroversial, as long as the project generates benefits. But even in this case there seem to be some hints that it is probable in ongoing intertemporal cooperation projects that knowledge could get lost. The knowledge that could get lost is the knowledge which is needed to control the project and its continuation.
3. Last but not least the problem of selection of scientific knowledge becomes the more urgent the more this knowledge is growing.

---

[11] Semantically stable means that information which has been transferred, protected from destruction and re-interpretation by decontextualisation etc. can be used to infer the intended relevant knowledge in a future context; comp. (Kornwachs 1995) p. 34.

[12] For a discussion of durability of data storage materials, copy processes etc. comp. (Kornwachs 1999b).

[13] The concept intertemporal cooperation ("intertemporale Kooperation") has been used by Dieter Birnbacher (1988). In presenting an utilitarian ethics of future Birnbacher points out that generation-overlapping projects are not a new but an old attribute of civilisations. They are started by one generation which expects and ensures by education that the following generations will continue the projects in their sense.

Ad 1) Since the early 80s the scenario "inadvertent human intrusion into the final repositories for high-level nuclear wastes" has been debated in the U.S.A. This scenario involves – according to risk studies – the highest risk for the integrity of the repositories. It has been argued in the following way: If knowledge about the repositories, their contents and the risks resulting from it would be available in a semantically stable way, future generations would not inadvertently intrude into the sites. This defines a "knowledge transfer problem".[14]

Closely connected to the dispute on repository systems there has been an intensive ethical debate. I suggest to use these arguments in order to define requirements for the so called information and documentation systems (IDS). Such systems would have the task to present the information which allows future generations to infer the knowledge needed for responsible actions at the disposal sites.[15] These ethical arguments could be used in a first order approach as arguments to justify norms for knowledge selection and transfer.[16]

"The polluter pays" principle demands the transfer of information about the final repositories if this promotes the repository performance. If the level of harms and risk should not exceed the level accepted in our generation and, as risk studies show, the risk of inadvertant human intrusion is the highest, and the risk of human intrusion correlates positively with growing ignorance then information and documentation systems seem to be obligatory. The obligation to open up options for responsible actions for future generations strengthens the claim for the transfer of the knowledge needed. The idea of "responsibility" seems to commit ourselves to communicate the reasons and backgrounds of the project. Therefore in the case of the final disposal of nuclear wastes the following norms of knowledge selection and transfer can be accepted as *prima facie* justified:[17]

1. **Norm (Warning):** Knowledge which is necessary to inform future generations of how to handle technical and cultural estates of our days needs to be transferred as warnings, use and maintenance advices and technical and organisational information.
2. **Norm (Explain):** It is obligatory to inform future generations about justifications, situations, institutions and persons with respect to intertemporal cooperation projects. One ought to explain the intertemporal cooperation project.

There is a considerable number of further cases in which one could argue in the same manner knowing the discussion about the transfer of knowledge in the waste isolation case. For example, often toxic chemical wastes get finally disposed close to the biosphere.[18] Or the "release of genetically engineered organisms" should be kept in mind, even if the risks connected to the release of these organisms are commonly estimated to be low.[19]

---

[14] Comp. (Human Interference Task Force 1984) and (OECD/NEA 1995b) p. 7.

[15] Comp. (Berndes, Kornwachs 1996).

[16] The arguments refered here origin from the OECD/NEA expert network.

[17] For the justification of the norms with arguments which are in discussion comp. (Berndes 1999).

[18] Comp. (Habeck-Tropfke, Habeck-Tropfke 1985) pp. 167.

[19] Comp. (Perrow 1992) pp. 342.

Ad 2) It is possible that in the future we will rely on infrastructures where the knowledge for their design and production has been lost. Huge software systems, information and communication structures and even automatized production systems based on software should be mentioned here. Probably computer programs will be used for a long time in which parts of it are reused.[20] Here claims for the documentation of software, its function and explanations about intended applications seem to me to be justifiable. Knowledge about software should be transferred which becomes increasingly an integral and safety-relevant part of our long-living infrastructures.[21]

Ad 3) The last approach starts from the anticipated growth of knowledge and science. Almost anybody knows about the thesis of an exponential growth of knowledge.[22] Also many people talk about decreasing half life of knowledge when they look at aging of technical qualifications, changing software paradigms, etc. And the competition among so called "orchidee-disciplines" at the universities for funds seem to become even stronger.

Knowledge needs for its preservation competent people who have acquired it for themselves. Even with respect to growing opportunities to improve the efficiency of knowledge transfer by communicating relevant information in a certain context (institution, organisation, artefacts) the transfer of the actual knowledge will come up against limiting economic factors.[23] Then criteria for the positive selection of knowledge should be considered. Of course, here ethical aspects should be taken into account.

4 About Justification of Norms of Knowledge Selection and Transfer

It is plausible to state with Julian Nida-Rümelin (1996) that ethical theories are normal theories. Hence scientific justification of descriptive and normative knowledge has the same holistic and coherentistic character. Then final justifications are impossible. Ethical knowledge is based neither on self-evident truths of reason nor on critique-resistant propositions which describe the conditions of the possibility of normative discourses. And ethical norms need not necessarily to be generalisations of our situated singular moral intuitions ("unserer situationsbezogenen singulären moralischen Intuitionen").[24] So far I have committed myself to a cognitive approach.

I also interpret moral propositions objectivistically.[25] The objectivistic alternative seems to me to be plausible, as long as our common moral language and the pragmatic of moral conflicts seem to suggest this choice ("moralische

---

[20] Comp. (Bullinger et.al. 1997) pp. 27.

[21] An example is NORAD, the central command of the north-american air defense; comp. (Perrow 1992) pp. 330.

[22] Comp. prominently (Solla-Price 1974) and (Rescher 1982).

[23] (Rescher 1982) argues with economic factors for his thesis of deceleration of knowledge production, his analysis implicates plausibly the thesis mentioned above; comp. also (Berndes 1999).

[24] Comp. (Nida-Rümelin 1996) p. 41.

[25] For a definition and critique of subjectivism and objectivism comp. (Kutschera 1999) pp. 59.

Alltagssprache und die Pragmatik moralischer Auseinandersetzungen eine objektivistische Interpretation nahelegen [...]").[26]

In the following, I will limit my considerations to Kantian deontology and the utilitarian approach.[27] Of course, there are many further paradigms of ethical justification like contract theory, virtue theory and eco-ethics but for today I would like to restrict myself to the mentioned approaches.

Norms of knowledge selection and transfer become introduced or woven into a net of already existing norms in order to justify them. In doing so moral beliefs are collected which are used or accepted in each of the fields of discussion.[28] Then the proposed norm candidates are tested for their "compatibility" with accepted ethical principles and norms from related areas. And it must be shown that if one follows the norms and realises the related rule of action given values get optimised.[29] A further argument for the validity of the norms results from the discussion of the coherence between related norms from related application areas (Bereichen)[30].

The principle of preservation of conditions (Prinzip der Bedingungserhaltung) is used to check the norms of knowledge selection and transfer for their compatibility with a modern ethical principle.[31] They should be revealed as consequences of the principle of preservation of conditions.

The mentioned cases, in which norms for the selection and transfer of knowledge should be applied, originate (in a broader sense) from ethics of technology. Therefore one should consult for the justification of the norms a catalogue of values which has been successfully used in this field. I have called the VDI-catalogue from the Richtlinie 3780.

In the discussion about the final disposal systems practical rules are widely used, which can be justified with relation to values and generalized situations.[32] Other methods like the cost-benefit analysis have not been very suitable.[33] In the context of more general considerations towards a rule utilitarian ethics of future Dieter Birnbacher (1988) suggests a set of so called praxisnorms which has been used also.

Last not least a list of duties has been collected for the justification of norms of knowledge selection and transfer. They origin from the nuclear waste case and modern (German) literature on deontology.[34]

---

[26] See (Nida-Rümelin 1996) p. 51.

[27] (Nida-Rümelin 1996) mentions further classes of normative ethics: libertarism, contractualism, ethics of virtue; comp. ibid. p. 44.

[28] Comp. (Berndes, Kornwachs 1996).

[29] In an act-utilitarian environment norms can only be motivated, if one argues from a rule-utilitarian point of view norms can be justified; comp. (Kutschera 1999) pp. 200.

[30] Nida-Rümelin claims that different normative criteria are relevant for different areas (Bereiche) of human practice which cannot be reduced to one single system of moral rules and principles; comp. (Nida-Rümelin 1996) p. 63.

[31] The principle of conservation of conditions is as follows: Act in a way that the conditions for potential responsible action will persist for all persons involved ("Handle so, daß die Bedingungen (der Möglichkeit) des verantwortlichen Handelns für alle Beteiligten erhalten bleiben"); see (Kornwachs 1999) p. 67.

[32] Comp. (Catron 1995) p. 132, (Birnbacher 1988) pp. 197.

[33] Comp. (Catron 1995) p. 129.

[34] E.g. (Gert 1983).

5   Norms of Knowledge Selection and Transfer

Further norms shall be presented in this section: All of them and the two norms presented above have passed the procedure which I have described.[35] Can we want – considering the actual rate of growth of knowledge – that a present future has to transfer all the knowledge of our days? Asking this way the answer is easy. Nobody can want that. Keeping knowledge present affords lots of resources, creativity and power which could be used somewhere else more fruitfully. Now one could argue that future generations always will have the opportunity to select the important knowledge from the irrelevant. But this means to leave the demanding selection for a younger inexperienced generation which only has unsure selection criteria at her hand. Therefore I propose the following norm:

3. **Norm (Deletion):** Everybody (either individual or organisation) who has knowledge and wants to pass on knowledge should filter and delete some information. The threat of an overflow of information needs the organisation of a process of disposal.

Of course, the actual generation has a certain interest in passing on its knowledge. The question once more considering the growing amount of knowledge is, in which manner the communication of knowledge should happen. The following has been proposed to clarify this:

4. **Norm (Offering/Recommendation):** It is necessary to offer the knowledge which is not obligatory to be passed by an appropriate presentation of information. This needs to be done in a way which does not imply the acceptance by the next generation.

Norms 3 and 4 require an active selection of knowledge and oblivion. An recommending or offering not forcing attitude is connected with them. This seems – as it can be clearly seen – not to be self-evident because knowledge is an emphatic concept. And scientific knowledge is always connected with persons and success of people in their social role. Last but not least there is a driving force for the actual generation from the hope that the own work which is embedded into a "context of sense" of an institution will be continued. This hope could reach up to a however secularised idea of immortality.[36]

6   Discussion of some Objections

In the case nuclear waste disposal 10 000 years have been considered as a suitable timeframe for the transfer of knowledge. Therefore even the first realization concepts of the Human Interference Task Force (HITF) have had pyramidal dimensions: Either semiotic – i.e. solutions which trust in the power of signs – or

---

[35] Comp. (Berndes 1999)
[36] Comp. (Kornwachs, Berndes 1998) pp. 21

institutional approaches are grandiose.[37] Sebeok e.g. proposed an "Atomic Priesthood" which would be in charge of the nuclear knowledge and would have to pass on the knowledge about the dangers and risks from the buried nuclear wastes.[38]

Of course these proposals have been criticized. Sebeok has been confronted with the critique that his idea of an atomic priesthood only articulates the fears of an U.S.-elite for their decline.[39] And another critique could ask for the value of knowledge and whether it would not improve the situation to forget most of our fateful knowledge.

Pictures of a destroyed world after a collapse of our civilisation and scenarios of happy people which do live in peace and freedom after a collapse have something in common: both are "pictures" of possible futures. "Pictures of the future" are very important to justify longterm actions. I hope that those who start or continue intertemporal cooperation have knowledge about the future, i.e. justified beliefs about their actions and effects and causal connections. This knowledge does also incorporate justified premises about future human beings with whom the intertemporal cooperation should be continued. It is possible to build a method on such knowledge which helps to take into account the interests of "future generations" without producing the problems commonly connected with future ethics:[40]

Method to solve the Limitation Problem of Future Ethics: Those who act are allowed to contribute only to futures (or: are allowed to start or continue longterm projects), in which their actions and their forseeable effects do not hurt any accepted (or probably accepted) norms (or values). That means that projects (or trajects) should be designed in a way that the situation for any group of imagined future people will not change in a way that for them the principle of preservation of conditions is hurt.

A premise of this method is that future human beings will be similar to us in their needs, desires, norms and values. Norms and values are universal.

Therefore we cannot desire for the generations to come another knowledge. Because to hope for that is as selfcontradictory as to really wish another ethics as the ethics we have accepted and justified. In the opposite, if we continue intertemporal cooperation we do have an interest that our future "cooperation partners" will understand us. Then one may imagine cultural discontinuities with massive losses of knowledge but this is not a part of an authoritative future for us. Then the following norm of knowledge selection and transfer is motivated:

5. **Norm (Transfer of the Knowledge in general):** The transfer of the knowledge in general (cultural context) towards an indefinite future is a necessary but not sufficient condition of the possibility to talk meaningfully about intertemporal cooperation projects and "Zukunft für uns".

---

[37] Comp. (Posner 1990)

[38] Comp. (Sebeok 1990)

[39] Comp. (Blonsky 1990)

[40] A problematic existence of non-existent subjects as suggested by (Birnbacher 1988) is from my point of view as unsatisfying as the ersatz of future-ethical considerations through an appropriate ("angemessene") representation of future generations by our children which has been proposed by (Ott 1997) p. 645

This is an answer to the question whether we should pass on our fateful knowledge. We should pass it on. And our task is also to learn lessons from our experience in applying the knowledge and to pass them on too.

To select with respect to a long timeframe and to transfer knowledge means to educate the following generation. I hope we will succeed in this task. The more we are able to answer honestly the questions of our children and grandchildren why we have implemented the one or the other technology and whether we also considered their interests the better we will succeed in education. Presumably our descendants then will have good motives to pass on the most important parts of our knowledge together with theirs.

If another society with knowledge different from ours comes into being – maybe after a catastrophe – and if this society discovers inadvertantly disposal sites, then I call it fate.

7   Summary and Application of the Norms

I have argued from a holistic/coherentistic concept of knowledge and presented, how oblivion, selection and transfer of knowledge could be understood. Normative knowledge and ethics as a academic discipline have been interpreted analogously. Ethical reasoning then is to order and systematize moral propositions.

I have understood ethics as cognitivistic and objectivistic. Accordingly norms of knowledge selection and transfer have been proposed and justifications for them have been sketched. Here I have used Kantian and utilitarian ethics. The resulting set of norms should be understood as a "prognosis" – these norms are valid for the "new" field of moral judgement. They have to prove themselves in praxis.

Finally I have discussed how ethical norms could be applied in problems which involve a long timeframe. The method to solve the limitation problem of future ethics demands that everybody who wants to start longterm actions has to evaluate them ethically in a way that considers those who he presupposes to be existent in his pictures of the future which he is usually using in order to justify his action.

There is a recommendation of five norms of knowledge selection and transfer. Of course, these norms need to be detailed. Particularly the norms define the task to develop preception criteria analogue to those which have been proposed by archivists.[41]

There are opportunities to apply the norms. The monitored-retrievable disposal option comprises of science and documentation centers. The existing museum of technology, e.g. the Smithsonian Air and Space Museum, and science centers are further institutional links, where selected scientific and technological knowledge could be handed down.

Requests to select knowledge and to delete texts can be understood as directed to the libraries. They have to decide on their purchase strategy, collection areas etc. with respect to growing numbers of publications and new options of information and communication technologies. And requests can be understood as

---

[41] Comp. (Lübbe 1992) pp. 191

directed to the scientific communities which can help by changing the incentives to decrease the number of publications and to select the knowledge of "lasting" interest.

References

Berndes S, Kornwachs K (1996) Transferring Knowledge About High-Level Waste Repositories. An Ethical Consideration. In: Proceedings of the 7th Annual International Conference on "High Level Radioactive Waste Management". Las Vegas. Nevada. 29.04. - 03.05.1996. S. 494 - 498

Berndes S (1999) Zukunft des Wissens – Vergessen, Löschen und Weitergeben. Ethische Normen der Wissensauswahl und –weitergabe. Entwurf der Dissertationsschrift. Lehrstuhl Technikphilosophie. BTU Cottbus. Juni

Birnbacher D (1988) Verantwortung für zukünftige Generationen. Reclam. Stuttgart

Blonsky M (1990) Wes Geistes Kind ist die Atomsemiotik? In: Posner (1990)

Bullinger H-J, Weisbecker A, Supe G, Frings S (1997) Software-Management komplexer Systeme. In: Bullinger H-J (Hrsg) Software-Technologien in der Praxis. Objektorientierung, Wiederverwendung, Componentware, verteilte Software-Architekturen. Fraunhofer. Stuttgart

Catron B L (1995) Balancing Risks and Benefits fairly across Generations: Cost/Benefit Considerations. In: OECD/NEA (Hrsg) Environmental and ethical aspects of long-lived radioactive nuclear waste disposal. Proceedings of an Int. Workshop. Paris. September 1-2. 1994. OECD. Paris. S. 129 - 141

Gert B (1983) Die moralischen Regeln. Eine neue rationale Begründung der Moral. Suhrkamp. Frankfurt/Main

Habeck-Tropfke H-H, Habeck-Tropfke L (1985) Müll- und Abfalltechnik. Werner. Düsseldorf

Hubig Chr (1993) Technik- und Wissenschaftsethik. Ein Leitfaden. Springer. Berlin u.a.

HUMAN INTERFERENCE TASK FORCE (1984) Reducing the Likelihood of Future Human Activities That Could Affect Geologic High-Level Waste Repositories. Technical Report prepared for the Office of Nuclear Waste Isolation. BMI/ONWI-537. Columbus OH

Kornwachs K (1995) Wissen für die Zukunft. Über die Frage, wie man Wissen für die Zukunft stabilisieren kann. Eine Problemskizze. BTU Cottbus. Fakultät 1. Bericht Nr. PT-01/1995. Cottbus

Kornwachs K, Berndes S (1998) Zukunft unseres Wissens. Ansätze zu einer Ethik intergenerationeller Kommunikationshandlungen. In: Forum der Forschung. Wissenschaftsmagazin der Brandenburgischen Technischen Universität Cottbus 4(1998)6 S. 19 – 25

Kornwachs K (1998) Von der Information zum Wissen? Alle wissen alles – keiner weiß Bescheid. Beitrag zur 120. Versammlung der Gesellschaft Deutscher Naturforscher und Ärzte "Informationswelt – unsere Welten der Information". Wissenschaftliche Verlagsgesellschaft

Kornwachs K (1999a) Bedingungen verantwortlichen Handelns. In: Timpe K P, Rötting M (Hrsg) Verantwortung und Führung in Mensch-Maschine-Systemen. 2. Berliner Kolloquium der Daimler-Benz-Stiftung. Pro Universitate. Sinzheim. S. 51 - 79

Kornwachs K (1999b) Haltbarkeit von Information und Tradierung von Wissen. In: Forum der Forschung. Wissenschaftsmagazin der Brandenburgischen Technischen Universität Cottbus 5(1999)9

Kutschera Fv (1999) Grundlagen der Ethik. de Gruyter. Berlin. New York 1999. 2. überarb. Aufl.

Lübbe H (1992) Im Zug der Zeit. Verkürzter Aufenthalt in der Gegenwart. Springer. Berlin u.a.

Luhmann N (1984) Organisation. In: Ritter J, Gründer K (Hrsg) Historisches Wörterbuch der Philosophie. Wissenschaftliche Buchgemeinschaft. Darmstadt. Bd. 6. S. 1326 – 1329

Mittelstraß J (1996) Wissen. In: Mittelstraß J (Hrsg) Enzyklopädie Philosophie und Wissenschaftstheorie. Bibliographisches Institut. Mannheim u.a. Bd 4. S. 717 - 719

OECD/NEA (1995a) (Hrsg) Environmental and ethical aspects of long-lived radioactive nuclear waste disposal. Proceedings of an Int. Workshop. Paris. September 1 – 2. 1994. OECD. Paris

OECD/NEA (1995b) (Hrsg) Future Human Actions at Disposal Sites. OECD. Paris

Perrow C (1992) Normale Katastrophen. Die unvermeidbaren Risiken der Großtechnik. Campus. Frankfurt/Main. New York. 2. Aufl.

Posner R (1990) (Hrsg) Warnungen an die ferne Zukunft. Atommüll als Kommunikationsproblem. Raben. München

Rescher N (1982) Wissenschaftlicher Fortschritt. Eine Studie über die Ökonomie der Forschung, Walter de Gruyter. Berlin. New York

Rohbeck J (1993) Technologische Urteilskraft. Zu einer Ethik technischen Handelns. Suhrkamp. Frankfurt/Main

Ropohl G (1996) Ethik und Technikbewertung. Suhrkamp. Frankfurt/Main

Schwemmer O (1984) Institution. In: Mittelstraß J (Hrsg) Enzyklopädie Philosophie und Wissenschaftstheorie. Bibliographisches Institut. Mannheim u.a. Bd. 2. S. 249 – 252

Sebeok T (1990) Die Büchse der Pandora und ihre Sicherung. Ein Relaissystem in der Obhut einer Atompriesterschaft. In: Posner (1990)

Solla Price D J de (1974) Little Science, Big Science. Von der Studierstube zur Großforschung. Suhrkamp. Frankfurt/Main

List of Authors

Banse, Gerhard, Professor Dr. sc. phil. (born in 1946), studied chemistry, biology, and pedagogics at Pädagogische Hochschule Potsdam, and philosophy at Humboldt Universität zu Berlin; 1969 diploma; 1974 dissertation and 1981 habilitation on philosophical problems of technology; 1974 - 1990 scientist and senior scientist at institute of philosophy of Akademie der Wissenschaften der DDR in the field of philosophy and history of technology; 1988 professor for philosophy; 1993 - 1998 senior scientist at department of philosophy of technology at Brandenburgische Technische Universität Cottbus in the field of philosophy of technology and general technology, lectures in the domains history and philosophy of technology, general risk research, technological change, theory of technical sciences and technology assessment (especially information and communication technologies, information technology security); 1997 - 1999 senior fellow at Europäische Akademie Bad Neuenahr-Ahrweiler in the field of technology assessment in Middle and Eastern European Countries; 1999 senior scientist and institute for philosophy at Universität Potsdam; since October 1999 scientist at the institute for technology assessment and systems analysis at Forschungszentrum Karlsruhe; guest professorship at the universities of Düsseldorf, Penn State (USA) and Banska Bystrica (Slovakia); professor of general technology at Brandenburgische Technische Universität Cottbus, and member of Leibniz-Societät Berlin. Postal address: Forschungszentrum Karlsruhe, Institut für Technikfolgenabschätzung und Systemanalyse, Postfach 3640, D-76021 Karlsruhe, Germany

Bechmann, Gotthard (born in 1945), studied law, political science, sociology and philosophy at the universities of Frankfurt/M and Berlin, diploma 1971, scientist at the University of Frankfurt/M 1973 - 1973 scientist Studiengruppe für Systemforschung Heidelberg, since senior scientist Forschungszentrum Karlsruhe, Institut für Technikfolgenabschätzung und Systemanalyse (ITAS), lecture at the University of Karlsruhe of risk research and technology evaluation, guest professorship at the universities of Bremen, Moskau (Russia), San Sebastian (Spain), Tampere (Finland); Member of the Board of the International Academy for Sustainable Development and Technology at the University of Karlsruhe; Co-editor of the year book "Technology and Society". Main areas of research and publication: technology assessment and risk research, sociology of science and technology, environmental research, environmental law and theory of society. Postal address: Forschungszentrum Karlsruhe GmbH, Institut für Technikfolgenabschätzung und Systemanalyse, Postfach 36 40, D-76021 Karlsruhe, Germany

Berndes, Stefan, Dipl.-Ing., (born in 1966), studied aerospace engineering, philosophy and political sciences at the University of Stuttgart, 1992 diploma, 1992 -

1995 research assistant at the Fraunhofer Institute for Industrial Engineering, Stuttgart, research and consultancy in questions of Concurrent/Simultaneous Engineering, 1995 - 1999 research assistant at the Brandenburg Technical University at Cottbus, research on philosophy of technology, since 1999 project manager, junior consultant of Prognos AG, Berlin. Postal address: Prognos AG, Dovestr. 2 - 4, D - 10587 Berlin, Germany

Efremenko, Dimitri V., Dr. phil. (born in 1967), studied History at the Lomonossow State University (Moscow), 1989 diploma, postgraduate studies at the Institute for Philosophy, Russian Academy of Sciences, and University of Karlsruhe (Germany), 2000 doctoral thesis on philosophical aspects of technology assessment, since 1997 scientist at the International Institute for Global Problems of Sustainable Development, International Independent University of Environmental & Political Sciences (Moscow). Main research areas: philosophy of technology, technology assessment, sustainable development in the countries in transition, international climate policy. Postal address: International Institute for Global Problems of Sustainable Development, International Independent University of Environmental & Political Sciences, p. b. 20, 111250, Moscow, Russian Federation

Filacek, Adolf, Dr. rer. nat. (born in 1944), studied mathematics (minor subject mathematical statistics and probability) at the Charles University in Prague, 1967 diploma, 1971 doctorate at the Charles University, 1980 Ph.D. at the Czechoslovak Academy of Sciences, 1971–1980 researcher at the Institute of Mathematics, 1981-1989 at the Institute of Philosophy and Sociology of the former Czechoslovak Academy of Sciences. Currently (since 1990) member of the scientific staff of the Centre of Science, Technology, Society Studies at the Institute of Philosophy and Secretary of the Humanities and Social Sciences Division at the Office of the Academy of Sciences. Main research areas: science and research policy, evaluation in science, statistical analysis of research systems in transition. Postal address: Centre of Science, Technology, Society Studies at the Institute of Philosophy, Academy of Sciences, Jilská 1, 110 00 Prague 1, Czech Republic.

Fobel, Pavel, doc., PhDr., CSc (born in 1953), head of the Chair on Ethics and Aestethics, Faculty of Humanities at the University of Matej Bel in Banska Bystrica, Slovakia. In 1977 he graduated from the University of Lomonosov, Faculty of Philosophy, specialisation at sociology. His doctorate and candidate works were focused onto methodological problems of a system analysis and its application on a social administration. He habilitated at Faculty of Philosophy, the University of Komensky, Bratislava in 1988. He initiated the conception of development concerning with departments of philosophy, ethics, aestethics and their particular specializations. He has supported and given lectures on History of Philosophy, Sociology, Social Philosophy, Ethics and Applied Ethics. He was the chairman of Philosophical Sciences and now he has been the head of the Chair on Ethics and Aestethics. He has supported the specialized study on applied ethics. He is a member of the Science Board at FHV UMB, the Science Board at UMB, the collective board for specialized doctorate study and a member of the editorial board at FU SAV. He is a doctorate trainer, organizer and supporter (garant) of international science symposia. He took an active part in the XXth World Congress of Philosophy, Boston, USA.

Gethmann, Carl Friedrich , Professor Dr. phil. habil., lic. phil. (born in 1944), Studies of philosophy at the universities of Bonn, Innsbruck and Bochum; 1968 lic. phil (Institutum Philosophicum Oenipontanum); 1971 Doctorate to Dr. phil (Ruhr-University of Bochum); 1978 Habilitation for „Philosophy" (University of Konstanz). 1968 Scientific assistant; 1972 Assistant professor for philosophy at the University of Essen; further lectureships at the universities of Düsseldorf and Göttingen. – Appointment to a full professorship (C 4) at the University of Oldenburg (1990), at the Center of Technology Assessment in Stuttgart (1991) as well as at the universities of Essen (1991), Konstanz (1993) and Bonn (1995); Director of the Europäische Akademie zur Erforschung von Folgen wissenschaftlichtechnischer Entwicklungen GmbH (1996); Member of the Academia Europaea (London); full member of the Berlin-Brandenburgische Akadamie der Wissenschaften. Main research areas: Philosophy of language/ Philosophy of logic, Phenomenology and Applied philosophy/ Technology assessment. Postal address: Europäische Akademie zur Erforschung von Folgen wissenschaftlich-technischer Entwicklungen Bad Neuenahr-Ahrweiler GmbH, Wilhelmstr. 56, D-53474 Bad Neuenahr-Ahrweiler, Germany

Gorokhov, Vitaly G., Dr. phil. habil. (born in 1947), studied electronic engineering at the Moscow Radio/Mechanical Technical School, 1965 diploma, philosophy at Moscow Lomonosow University, 1971 diploma, post graduate studies of philosophy of science and systems theory at the Institute for the History of Science and Technology of the Soviet Academy of Sciences (today - Russian Academy of Sciences - RAS), 1975 - received his doctorate dissertation on methodology of systems engineering, 1986 - received his habilitation dissertation on methodological analysis of the development of the theoretical knowledge in the modern engineering sciences, both in the Institute for Philosophy of the RAS; 1971-1977 - Research Institute for Radiocommunication and Control, chief of department for the methodology of systems engineering, 1977-1989 - as a member of the editorial staff of Voprossi filisofii („The Problems of Philosophy") - the leading philosophy journal in the USSR, headed a department for philosophical and social problems of science and technology; 1978-1989 - reader and from 1985 professor for the philosophy of science and technology at the Moscow Mining Institute (Technological University); since 1988 - Institute of Philosophy of the RAS, a Philosophy of Technology Research Group, head and since 1992 - leading scientist; since 1994 - deputy director of the International Institute for the Global Problems of the Sustainable Development of the International University of the Ecology and Politology (MNEPU) in Moscow; scientific coordinator of the German-Russian Postgraduate College (since 1995) and the International Academy for Sustainable Development and Technologies (since 1999) of the University of Karlsruhe; since 1995 - professor and head of the department for the philosophy of science and technology in the State University for Human Sciences (GUGN) in Moscow. Main research area: philosophy of science, technology and environment, global problem of the sustainable development, methodology of systems engineering and project management. Postal address: University of Karlsruhe, Kollegium am Schloss II, D-761128 Karlsruhe, Germany

Grunwald, Armin, Professor Dr. rer. nat. (born in 1960), studied physics at the universities Münster and Cologne, 1984 diploma, 1987 dissertation on thermal transport processes in semiconductors at Cologne university, 1987-1991 systems specialist, studies of mathematics and philosophy at Cologne university, 1992 graduate (Staatsexamen), 1991-1995 scientist at the DLR (German Aerospace Center) in the field of technology assessment, since 1996 vice director of the European Academy, 1998 habilitation at the faculty of social sciences and philosophy at Marburg university with a study on culturalistic planning theory. Since October 1999 director of the institute for technology assessment and systems analysis (ITAS) at the research center Karlsruhe. Postal address: Institut für Technikfolgenabschätzung und Systemanalyse, Forschungszentrum Karlsruhe, Postfach 36 40, D-76021 Karlsruhe, Germany

Kiepas, Andrzej, Professor Dr. phil. habil. (born in 1950), studied physics at the University in Katowice, 1975 diploma, 1982 doctorate at the University of Silesia in Katowice on the problems of technology assessment, 1991 habilitation at the Technological University in Dresden on the problems of responsibility in technological development. Since 1975 assistant of the Institute of Philosophy at the University of Silesia in Katowice and since 1995 professor at the same university. Since October 1999 director of the Institute of Philosophy at the University of Silesia. Postal address: Institute of Philosophy, University of Silesia, Bankowa 11, PL-40-007 Katowice, Poland

Kornwachs, Klaus, Prof. Dr. phil., studied physics, mathematics and philosophy at the Universities of Tübingen, Freiburg, Kaiserslautern and Amherst (MA). 1973 Diploma degree in Physics, Dissertation 1976 (Philosophy of Language) and Habilitation 1987 (System Theory). Lecturer at University of Freiburg, 1977-198 Research Assistent at the Institute for Border Areas of Psychology, Freiburg, 1979-1992 Research Fellow at the Fraunhofer-Society for Applied Science, Stuttgart, since 1989 Director of the Department for Personal Management and Technology Assessment. Since 1990 Honorary Professor for Philosophy at University Ulm, since 1992 Chair for Philosophy of Technology at Brandenburg Technical University of Cottbus (BTUC). Founder and President of the German Society for System Research; 1991 SEL-Award for Communication Technologies. Director of the Center for Technology and Society at BTUC. Postal address: Lehrstuhl für Technikphilosophie, BTU Cottbus, Karl Marx Strasse 17, D - 03044 Cottbus, Germany

Langenbach, Christian J., Dr.-Ing. (born in 1963), studied Aerospace Technology (Main field: New Technologies) at the Universities of Munich and Stuttgart. 1991

Bachelor of Engineering, 1997 doctoral thesis on a theoretic-analytical work about parametric comparison of space transportation systems in particular consideration of state of engineering modelling at the Technical University of Berlin, 1991 to 1996 member of research staff of Space Systems Analysis Division at German Aerospace Research Establishment (DLR), since April 1996 member of scientific staff at Europäische Akademie GmbH, Projectmanager of the project group "Electronic Signatures. Cultural rules and moral responsibility", responsible person for the IT-sector. Main research areas: Technology Assessment, Technology Management, Information Society. Postal address: Europäische Akademie zur Erforschung von Folgen wissenschaftlich-technischer Entwicklungen Bad Neuenahr-Ahrweiler GmbH, Wilhelmstr. 56, D-53474 Bad Neuenahr-Ahrweiler, Germany

Machleidt, Petr, Dr. phil. (born in 1949), studied sociology and economy at the Faculty of Philosophy, Charles University in Prague, 1974 diploma, 1976 doctorate at the Charles University. From 1974 member of research staff (in Academy of Sciences) in the field of science, technology, society relations, 1975-1990 at the Institute of Philosophy and Sociology of the former Czechoslovak Academy of Sciences. Currently (since 1990) member of the research team of the Centre of Science, Technology, Society Studies at the Institute of Philosophy of the Academy of Sciences, Czech Republic. Main research areas: Social and human dimensions of science and research assessment, Transformation of science systems, Technology Assessment, Comparison of TA-Concepts. Postal address: Centre of Science, Technology, Society Studies at the Institute of Philosophy, Academy of Sciences, Jilská 1, 110 00 Prague 1, Czech Republic

Okoń-Horodyńska, Ewa, Professor Dr hab. (born in 1950), studied Economics at Academy of Economics in Katowice, 1974 diploma, 1985 doctorate on restructurization of Polish enterprises at Academy of Economics awarded by Ministry of Labour and Social Policy, 1990 habilitation at Academy of Economics in Katowice, 1990-1991 International Exchange Holland Program Fellowship at the Netherland Economic Institute and Erasmus University in Rotterdam, 1992 visiting professor Fellowship at Brown University in Providence, at the University of Washington, Henry Jackson School of International Studies in Seattle, and at George Washington University in Washington D.C., IREX Programme; 1997 Training Awards and Institutional Links. British Council Research programme on British Science & Technology Parks, membership of professional bodies: Polish Economic Association, Warsaw and European Association for Evolutionary Political Economy, Oxford; 1991-1993 vice director of Economies Institute at Academy of Economics in Katowice, since 1996 Head of Global Economics Unit, since 1995 a Member of Scientific Board at the Centre for Industrial Management and at Institute of Management, Jagiellonian University in Kracow, since 1999 consultant at Ministry of the Economy, since March 1997 rector of Higher School of Management and Social Sciences in Tychy. Research area: Institutional Economics, Economics of Technical Change and Theory of Innovation, lecturer on Economics and Theory of Innovation. Postal address: Higher School of Management and Social Sciences, Plac Św.Anny, PL-43-100 Tychy, Poland

Provazník, Stanislav, Dr. phil. (born in 1932), studied philosophy and history at the Faculty of Philosophy, Charles University in Prague, 1955 diploma, 1966 Ph.D. dissertation at the Czechoslovak Academy of Sciences (on sociological problems of interests). From 1969 member of scientific staff (in Academy of Sciences) in the field of science, technology, society relations, 1970-1990 at the Institute of Philosophy and Sociology of the former Czechoslovak Academy of Sciences. Currently (since 1990) head of the research team of the Centre of Science, Technology, Society Studies at the Institute of Philosophy and member of the Economy Council at the Office of the Academy of Sciences. Main research areas: social and human dimensions of science and research assessment, transformation of science systems, science and research policy. Postal address: Centre of Science, Technology, Society Studies at the Institute of Philosophy, Academy of Sciences, Jilská 1, 110 00 Prague 1, Czech Republic

Salomon, Jean-Jacques, Professor, Dr. (born in 1929), honorary professor of Technology and Society and director of the Research Center Science, Technology and Society at the Conservatoire National des Arts et Métiers (CNAM), Paris, France. He was graduated from the Sorbonne in philosophy and anthropology, where he obtained his PhD in philosophy and history of science on the role od the scientists in the modern world. From 1963 to 1983 he created and directed the Science and Technology Policy Division of the Organisation for Economic Cooperation and Development (OECD). He has founded and chaired the International Council for Science Policy Studies attached to the ICSU network, presided the Standing Committee for Social Sciences of the European Science Foundation (Strasbourg) and more recently the Collège de la Prévention des Risques Technologiques attached to the Prime Minister offfice in France . He was invited professor at MIT, Harvard, Montreal University, Sao Paulo Institute for Advanced Study and is fellow of Clare Hall, Cambridge, UK, and member of the New York Academy of Sciences. Among his many publications and contributions, his major books are Science and Politics (MIT Press, 1973), Mirages of development (Lynne Rienner, Boulron/London, 1989), Science, War and Peace (Economica, Paris, 1989), The Uncertain Quest (UNU, Tokyo-Paris-NewYork) Le destin technologique (Balland/Gallimard, 1991) and Survivre à la science: Une certaine idée du futur (Albin Michel, Paris, 1999). Postal address: Centre Science, Technologie et Société, Conservatoire National des Arts et Métiers, 2, rue Conté – 75141 Paris, France

Schienstock, Gerd, Professor, Dr. rer.pol. (born in 1942), studied economics and sociology at the Technical University of Berlin, 1969 diploma, 1974 doctorate at the Technical University of Berlin, 1972-1974 Researcher at the Institute for Sociology at the Technical University in Berlin, 1974-1980 Assistant Professor at the University in Hamburg (Industrial Sociology), 1980 Habilitation at the University of Hamburg with a study on industrial relations, Germany, 1980-1992 Head of the Department of Sociology at the Institute for Advanced Studies in Vienna, Austria, 1993-1995 Senior Researcher at the Center for Technology Assessment in Baden-Württemberg, since 1995 Research Professor and Scientific Director of the Work Research Centre at the University of Tampere, Finland. His main research areas are innovation systems, information society, technological change and work organisation. Postal address: Work Research Centre, FIN-33014 University of Tampere, Finland

Tondl, Ladislav, Professor Dipl. Ing. (born in 1924), studied philosophy, sociology and economics, 1949 diploma, 1949 doctorate at the Charles University, 1953 habilitation in the field of philosophy, 1956 Ph.D. and 1966 D. Sc. at Czechoslovak Academy of Sciences. From 1959 to 1968 in a department in the Institute of Information Theory and Automation of the former Czechoslovak Academy of Sciences. In 1968 professor of philosophy at the Faculty of Philosophy at the Charles University in Prague, the same year had founded the Centre for Science and Science Policy Research at the former Czech. Acad. of Sci., in 1971 had to leave the Centre and scientific position from political reasons, till 1989 was not allowed to publish any papers in humanities and philosophy. In 1990-93 director of the Institute of Theory and History of Science of the Academy of Sciences, currently (since 1993) head of the Centre of Science, Technology, Society Studies. Main research areas: semantics and semiotics, philosophy of science and technology, information dimensions of technology, about 200 papers, more than 10 books. Postal address: Centre of Science, Technology, Society Studies at the Institute of Philosophy, Academy of Sciences, Jilská 1, 110 00 Prague 1, Czech Republic

Ulrich, Otto, Regierungsdirektor Dr. rer. pol. Dipl. phys. (born in 1942), studied physics at the Fachhochschule Iserlohn, 1964 diploma, 1975 educational economy M.A. at the Technische Universität Berlin, 1978 doctorate at the Freie Universität Berlin, worked in the industry, the Federal Chancellery and the German Bundestag, at present Bundesamt für Sicherheit in der Informationstechnik, head of department for technology assessment. Main research areas: politically control of technology, future of society and interculturality .Postal adress: Bundesamt für Sicherheit in der Informationstechnik (BSI), Referat II 5, P.Box 20 03 63, D-53133 Bonn, Germany

Zacher, Lech W., Professor of Sociology (born in 1942), studied political economics at the Warsaw University, 1965 M.A. diploma, Ph.D. in economics in 1971 (from Polish Academy of Sciences) and habilitation in 1977 (in sociology) also from the Academy. Worked in Warsaw University (1965-68); Polish Academy of Sciences (1970-1987); University of Maria Curie-Skłodowska (in Lublin, 1987-1995); also in the Prime Minister Office (1991-92); Military Academy of Technology (1992-1995); Academy of Entrepreneurship and Management, Warsaw (1995-present); University of Silesia, Dept. of Radio and TV (Katowice, 1996-present); Government Center for Strategic Studies (1998-present); Currently

Head of the Chair of Social Sciences at Leon Kozminski Academy of Entrepreneurship and Management in Warsaw, also the Head of the Center for Study of Information Society at the University of Silesia (in Katowice). He is also the advisor to the President of the Governement Center for Strategic Studies. Main research areas: Projects: on Technology, Democracy and Environment; on Technology, Power and Conflicts; future studies (information society); political philosophy, global problems and transformations (especially in Eastern Europe). Postal address: LKAEM, Jagiellonska 59, PL-03-301 Warsaw, Poland

Zahradník, Rudolf, Professor (born in 1928), physical chemist, Professor of Charles University in Prague 1967-, studied at the Institute of Chemical Technology in Prague, Dipl. Ing. 1952, Ph.D. 1956, D.Sc. 1968, Director of the J. Heyrovský Institute of Physical Chemistry ASCR 1990-93, President of the Academy of Sciences CR 1992-, President of the Czech Learned Society 1994-97. Main research areas: theoretical and applied quantum chemistry, molecular spectroscopy, chemical reactivity and weak intermolecular reactions. From 1965 to 1990 visiting professor at twelve universities in Europe, the U.S.A. and Japan. Many honorary degrees and memberships at universities and scientific societies. Postal address: Academy of Sciences of the Czech Republic, 117 20 Prague 1, Národní 3, Czech Republic

Printing: Mercedesdruck, Berlin
Binding: Buchbinderei Lüderitz & Bauer, Berlin

3876748

DATE DUE